Crimson Code

Crimson Code

The Price of Success

Desmond Jackson & Christopher Love

Evermore Books

Copyright © 2024 by Desmond Jackson and Christopher Love

All rights reserved. No part of this book may be reproduced in any manner whatsoever without written permission except in the case of brief quotations embodied in critical articles and reviews.

First Printing, 2024

Cover design by Jared Helton

This book is dedicated to Dr. Katie Welch Jackson

1

Preface

Desmond Jackson burst into my office one day in the spring of 2016, spouting something about a police raid on his dorm room. I didn't follow his rush of words, but the fear and shock I saw in his face told me this was serious.

Each class brings with it a new cast of characters in which some students immediately stand out—for various reasons. Desmond was one of those standout students. He asked questions, contributed to class discussion, and emailed me throughout the semester to make sure that he was doing the assignments correctly. One email in particular comes to mind in light of later events: he wanted to make sure he had not plagiarized and that he had correctly documented his sources. He was the kind of dedicated, ethical student that makes a teacher's job feel meaningful.

We discussed assignments after class, and in the course of our talks, we laughed and joked, and began to develop a rapport, a mutual respect and trust. That, I think, is why he believed he could come to me the day the troubles began. As those troubles escalated, as he was being threatened with expulsion and jail, I wrote a letter to the University to assure it that Desmond was the kind of student the University of Alabama should be proud to have among its student body.

While his legal saga unfolded over the next couple of years, Desmond and I would meet, have lunch, and discuss what he was going through, and the thought of writing a book about it all began to take form.

But I soon realized that there was a deeper story behind the legal story. Where had this young Black man acquired the resolve to stand up to a global corporation and its powerful law firm? In our talks, Desmond had told me a bit about his grandfather and the town he was from, and I suspected that the answer lay there.

That led me to travel to Lowndes County, Alabama, the "Bloody Lowndes" that stood at the forefront of the struggle for voting rights in the South in the 1960s. Desmond arranged for his grandfather, John Jackson, to meet me.

My journey took me down Highway 82, which cuts deep into the rural farmland of Alabama, and for me, deep into history. As I passed through Selma and crossed the famous Edmund Pettus bridge, I felt myself in the grip of that history. And when I met Desmond's grandparents and learned more of their role in that struggle, I knew that this history was a critical part of Desmond's story.

Though it is a collaboration, this book is indeed Desmond's story. These events happened to him. And as we worked on it, Desmond shared with me how important the book is to him. It would bring him peace of mind, allow him to express his frustrations, and satisfy his desire to get the truth out. What happened to Desmond mattered, and could happen to anyone—including anyone reading this book. And so we knew that we had to tell his story.

-Christopher Love

2

Prologue

Prologue
September 18, 2018. The law office of Huie, Fernambucq, and Stewart, LLP. Birmingham, AL. Mr. Tankersley and Mr. Taunton are seated across from me. They are with Balch & Bingham, LLP, also a Birmingham firm. Others are present, including my attorney, Mr. Daniel Fortune. But Mr. Tankersley, from Balch & Bingham, rises. He will take the lead in questioning me.

This is the man who sent me threatening emails and told my mother over the phone that I was going to prison. His face is long, smooth, and bone white, except for flushes of rose on his cheeks. The folds over his eyes, circled by his black-rimmed glasses, give him a narrow, suspicious look to counter his otherwise boyish face. A paisley handkerchief puffs from his blue pinstriped suit. Above his blank forehead his slick-backed hair lies in shallow waves. He's maybe 5'3 with an inch coming from the heels of his gleaming black shoes.

"Mr. Jackson," Tankersley begins, "would you please state your full name for the record?" His voice has that air of forced politeness that goes along with the rich Southern drawl, tinged with the familiar condescension of wealthy Southern whiteness. Tankersley wants to get me. He's been trying to get me for three years, and I think I see a mixture of frustration and satisfaction in his demeanor as if he's merely won a consolation prize. Still, I'm in front of him now, a twenty-one-year-old black student, and I think he's hoping to salvage the game and still get a piece of me. He doesn't seem to believe that I am adept enough to escape.

I tell him: "Desmond Armani Jackson."

Tankersley, though, can't wait to speak. He jumps in: "My name is Will Hill Tankersley." He states it so quickly it's almost an interruption. His abruptness and the singsong sound of his name amuses me, but I suppress my smile. Everything is for the record.

He continues with the confidence of a seasoned attorney: "In the course of this deposition I'm going to try to keep within the white lines here; if something comes up where I unintentionally stray into an area where you're uncomfortable or you think you need guidance, it's a right that you have to confer with your lawyer." Going over guidelines, he tells me I can take a break if I need to. "This isn't supposed to be a death march," he adds grimly.

I've already been on a death march for two years. He and his client have dragged me on it. This man, who now feigns politeness, tried to see to it that I went to jail and accused me of breaking federal and state laws. The Federal Computer Fraud and Abuse Act. The Federal Defend Trade Secrets

Act. The Alabama Trade Secrets Act. Trespassing. Unjust Enrichment. He wanted to put me away.

At the time of my interrogation, I was a senior at the University of Alabama. While in school, I'd started my own cybersecurity company, Jackson CS Consulting LLC, along with earning an internship at Google. I'd gotten interested in technology more than a decade earlier, when my grandmother talked my mom into getting her bored eight-year-old kid a computer. I never could have dreamed that it would lead that eight-year-old kid from one of the poorest counties in Alabama to interning at one of the most powerful internet companies in the world. Nor could I have known that it would lead me to a face-off with a lawyer who represented one of the largest publishing companies in the world. Yet here I was, in a conference room of a law firm, surrounded by attorneys who wanted to know how I did what they and a team of their client's experts couldn't figure out.

When I entered the University of Alabama in August 2015, I planned to make the most of my time. I was already fast acquiring expertise in cybersecurity, and I'd begun developing my own software. I was on my way to a career in tech, but I wasn't going to drop out of college, like some famous tech entrepreneurs. That was not an option, not in my family.

I came from a family who had long valued education. My great-grandfather and great-grandmother, Mathew and Emma Jackson, grew up in rural Alabama, working the cotton fields. They raised ten children and made damn sure that every one of their children understood the importance of an education. My grandfather and great-aunts and uncles all ended up going to college. This was a rare thing in Alabama in those days, even for whites, but my family believed that an education was the best defense against racism, political corruption, and poverty. Coming from a cotton plantation, my family understood that education was vital to building a better life. They'd lived through and participated in the Civil Rights Movement and were among the first blacks to register to vote in southern Alabama during the 1960s. I grew up knowing that education provides options and opportunities, and is key to the American Dream.

So there was no way I was going to bypass college.

Besides, I realized I still had much to learn. A formal education in computers could only help me diversify my skills and knowledge, I realized, and professors at the University could teach me more about cybersecurity and computer science. Then too, I looked forward to making contacts and networking with professors and students as I began to strategize long-term business plans and career goals. I imagined opportunities like conferences, job fairs, and internships that would put me in contact with experts and professionals from all over the country.

My idea of what I could get out of college may not have been the same as my parents', but we were pushing in the same direction.

If you get the idea I was ambitious, you're right, and I was focused. At UA, I wanted to get an early start and not wait until I graduated to jump-start my future. I had already been employed by software companies while I was in high school, and I knew that developing an innovative and aggressive business plan was crucial to my success in the fast-paced, ultra-competitive world of digital technology. As fast as the tech world changes, I didn't want to waste time by failing to take advantage of all opportunities. Get my name and work out there, I told myself, and I could attract those big talent-hungry companies, those cutting edge startups, or lesser-known but high-profit organizations that value creativity and initiative. That was my plan.

I wanted what many people dream about—quick success—and I believed I could get it. I would get it by creating software that lots of people just had to have.

In December 2015, while staying at my grandparents' house in the country near where my family once worked those cotton fields, I stumbled upon something that would leave a billion-dollar-corporation, the largest education publishing company in the world, trembling in fear.

By the following spring, I'd be detained by police, my computers seized, and be facing expulsion from the University of Alabama. No professor could teach me the lessons I was about to learn. I'd set out to create an innovative yet accessible piece of software. In doing so, I'd hoped to attract companies that would offer me a career or that would want to buy or license my product. Instead, I attracted enemies.

Back inside the law firm, in the stuffy atmosphere of that conference room of Huie, Fernambucq, and Stewart, Mr. Tankersley shifts gears. He's ready to get into it. He asks me about my current employment at Synopsys, a global tech and internet company. It's as if he's trying to catch me in a lie or a truth that he can seize to his advantage.

"What are you doing for them? For Synopsys?" he demands.

I'm hesitant but offer a reply: "I'm in the R&D department, research and development. I create cybersecurity tools and pen-tests on the side." I'm not sure if he knows what any of that means. But I figure that's his problem. His next question leads me to suspect his ignorance.

"Beyond what you've said. What are you doing for them, beyond R&D and cybersecurity?"

I don't like where he's going. Neither does my attorney, Daniel Fortune. He objects. "I think he's done enough of that."

Tankersley is visibly taken aback. I know why: he thinks that because he's an attorney he has a right to know everything about everybody else. But this man doesn't know me, nor does he even care to really know me. He asks me if I'm declining to answer the question, implying that my attorney and I are being evasive, when, really, it's none of his business. He alters tactics, trying a soft approach, and asks me about my educational plans. Then, he digs into what he really wants to know: "How did you become competent in computers?"

Mr. Fortune interrupts again to stop that line of questioning. "Counsel, if you'll just limit to what he did with your client."

Tankersley gets the picture; Fortune isn't going to allow him to push me around.

"Ok," he concedes. "Let's just start from the beginning."

CHAPTER 1

The Raid

The University of Alabama is located on a rocky bluff south of the Black Warrior River, which twists and turns southwesterly across the state. The campus sprawls across the northern part of Tuscaloosa, and in the spring, the grounds are awash in fresh green grass and trees flush with leaves that shade its walkways for the close to 40,000 students who traverse campus, shuffle among classes, meet up with friends, or head wherever their lives are taking them. Young, hopeful students like me, with varying degrees of naiveté and savvy, crisscross the Quad. UA is much like any other large state university: it's not just tailgating and football games at Bryant-Denny Stadium.

When I entered the University of Alabama in fall 2015, I already saw myself as working on my career. My mom wanted me to get a job so I could pay for school and living expenses, but I wasn't interested in the usual student jobs. I wanted to see if I could put my computer skills to work and get a gig online or work remotely. Though UA is steeped in tradition, I was hoping to meet students or professors who could clue me in to hidden opportunities. I was looking for something nontraditional, some innovative way to make money and grow my skills. I pushed myself not to let reticence hold me back or keep me from learning everything I could.

One question I had always learned to ask myself was "How does this work?" It's how I learned so much about computers so quickly. I entered college with many of the same problems as other students. I needed money, I needed to learn to manage my time, and I needed to make sure that I didn't get behind in my classes while trying to have a social life. So I approached college like I approached computers: I sought to figure out how it worked. To me that meant figuring how UA could serve my immediate career goals. One thing I recognized immediately was a ready market. I saw the student body as potential customers, customers of a business that, admittedly, I hadn't yet built. But before I could build it, I needed to know how Alabama students worked: not individually and not just some of them, but all of the student body collectively. I tried to imagine what all UA students needed or wanted or were interested in.

I sure didn't want to make the mistake of thinking of UA as just a football school because I'd miss the finer details of how the school worked and what kinds of people make up its student body. Including someone like me.

I never much cared for sports, so I was not the stereotypical Alabama student. I didn't get caught up in the Bacchanalian celebration of Saturday game days in the fall. Two years earlier in 2013, I was attending the Alabama School of Mathematics and Science, a selective high school in Mobile, when I got an opportunity to come to a summer educational camp at the University. It was a great opportunity, but while there, I shut down the campus's wireless network for five minutes. To impress a girl. Not smart. Well, smart, but not wise.

It didn't impress the university. I had pulled pranks like that for years, like manipulating online games like Call of Duty and World of Warcraft. At least my juvenile self thought of them as pranks. I wasn't mature enough to understand the ethics, morality, or even legality of what I was doing, but that was all soon to change. UA wasn't amused by my latest prank. Cameron Purvis, a Senior Information Security Analyst at UA and later a mentor of mine, let me off with a warning, and it changed my perspective. I started thinking about using my talents and skills in more positive, productive — and entrepreneurial — ways.

A couple of years later I became a student at the University. It was expected of me. My mother, Nina Jackson, graduated from Alabama, and it's where she met my biological father.

One of my first mentors at UA, Dr. Albert Lilly from the Alabama School of Mathematics and Science, helped convince Dr. Jeff Gray at UA that I had real potential as a student and as a code writer. Dr. Lilly taught at ASMS for twenty-three years before retiring in 2014. I was saddened to learn that he passed away three years after he retired. He was one of the first teachers to believe in me, and he was a factor in helping me realize the productive and positive aspects of my abilities. Rather than use my skills and knowledge for my own amusement, I could use them to build and contribute to computer science. As well as — and this was important to me — learning how to build my own business.

My first year at Alabama went well. I was ingratiating myself into the student body, and I met my girlfriend who was studying to be an electrical engineer. I also started my own company. I had already started to professionalize myself as an expert in cybersecurity, having worked for Enjin and Nitrogen Sports. By the spring of 2016, I was making a name for myself while gaining confidence and experience. I felt I was living up to Dr. Lilly's expectations and proving that he was right in believing in me.

Men with Guns

Then, in the early afternoon of April 25th, 2016, all that changed. I got a call from my best friend, Anfernee Wallace, who said, "Des, you gotta get over here." The "here" was my dorm room located in Lakeside West, a residential building on campus. I'd been sleeping in at my girlfriend's dorm in Riverside North, just across the street, and when Anfernee called, I was still half-asleep.

"What's going on?" I asked.

"The police, man. They're all over the place." "What do you mean? At my place?"

"Yeah. They came storming in."

I didn't believe him. "Man, I ain't got time for this," I remember saying. "Desmond, I'm serious. You gotta get over here."

I had a vague idea what this was all about, and I figured it was just a matter of me walking over to Lakeside West and straightening up the misunderstanding with the cops and everything would be alright. I had yet to fully understand what happened and its ramifications. But I can't say I didn't have a sense of foreboding.

Growing up black in America, I had to deal with friends of mine being shot and killed. When I was in fourth grade, my best friend, Monte, hanged himself. It's a statistical reality: young black men like me die at a rate out of proportion to the rest of the country, whether it's through violence or suicide. It's one of the reasons I like to ask everyone I know how they're doing. You never know what's going on in their lives, and you never know what may happen to them from one day to the next. Those high-profile cases of young black men killed in confrontations or encounters with police, in addition to what I already knew and grew up with, weighs heavily on my mind, as they do with so many other black people, especially young black men, especially in the South.

So when I say I wanted to understand how life at UA worked, this was a part of it. Although the University community is diverse and the culture more complicated than it may seem, there are still statistical and historical realities that confirm common perceptions. Blacks make up more than 26% of the population of the state of Alabama, but only about 10% of the student body at the University. It's not uncommon for a black student to be the only black person in several of their classes. I don't know why there is such a disparity, but I do know that some black Alabama parents are hesitant to encourage their non-athlete kids to go there because of their impressions and memories or memories of their parents of the University's ugly, racist history. Not only does this pertain to the era of segregation, but also because of more recent racist events that have happened on campus.

In 2013, the school newspaper, The Crimson White, uncovered racial discrimination in the University's Panhellenic sororities, including an incident when a highly qualified African American student was rejected for admission because of racist alumni. There was the time in 2017 that a student group intended to invite the noted white supremacist Jared Taylor to speak on campus. Fortunately, because of protests from students and faculty, his invitation was rescinded. If you consider that black students only make up about a tenth of the student population, these incidents can have compounding, detrimental, threatening, and alienating effects on them. It's safe to conclude that black families from around the country are equally if not more hesitant to encourage their children to attend the school. This certainly accounts for at least part of the low enrollment of African Americans at UA.

To me, these realities were personal, but I was determined to see them through the eyes of an entrepreneur. Perceptive entrepreneurs don't just accept social realities; they engage with them, and in turn, they help shape new realities and possibilities. That's why I was determined to understand the demographics and culture of UA. As I saw it, one of the reasons the demand for digital technology has exploded over the past decades is because companies like Apple created products that transcended race, gender, and nationality. Success grows exponentially when you plant its seed in as many fertile fields as possible. What is it that everyone needs? How do I accommodate them? These were the questions that motivated me, and that drove me to understand the perceptions and values of others. I wanted to know how students felt and experienced campus life. And I wanted to see how the administration reacted to events, because administrative reactions tell you how well organizations are run. Although the UA administration quelled racial incidents on campus, it seemed to me that the University was always reacting rather than being proactive. I didn't see the administration as working actively to create an atmosphere where all students felt welcome.

All of this was in my mind as I walked across the street to Lakeside West to meet with the police, and I can't deny that my murdered friends, the suicides, the national racial incidents with police and those on campus, gave me a sense of foreboding. I tried to ignore all that and to focus on explaining myself as professionally as possible. As far as I was concerned, this was a business matter and not a matter in which the police needed to get involved.

Because I wasn't just a student, I was an entrepreneur. After all my observations, socializing, and networking, the puzzling over what business I wanted to start, an opportunity arose, and I took advantage of it. Opportunities often come in the most surprising ways, I'd had it ingrained in me, so it's important to keep your eyes open and act when the opportunity presents itself. That's what I did, and as it turned out, that's what led the police to burst through my door.

Any thoughts that the police and I were going to have a cordial conversation over what they would come to understand as a simple misunderstanding vanished the moment I entered my dorm. Ten to twelve officers had swarmed and guarded my room, some of them hulking guys who looked like they could've been SWAT team members. Anfernee's warning that cops had come storming in wasn't an overstatement: that was exactly what had happened. In addition to University of Alabama police, cops from the Tuscaloosa Police Department were brought in as backup.

Backup for what? I certainly hadn't done anything violent. Nor had I threatened violence. What had I done? I had studied Pearson's MathLab, a math study aid where math students do homework exercises outside of class, and I had written a program that let students see the answers to the exercises. The UA police and the Tuscaloosa police need to storm into the dorm room over that? What was especially scary was that only my friend Anfernee and his girlfriend were there, and they had no idea what was going on. How was that safe for anyone involved? All that raced through my mind as I stood there surrounded by enough law enforcement power to take down a serial killer.

Later, I realized that, in order to stage this raid on my dorm room, the police had had to go to a judge and get a warrant. By that point, the police would have known that no violence was involved. They would have known they were going into a college residential building where dozens of students lived in proximity. They would have known they didn't need to escalate the situation. The question has to be asked: if I had been a white student would they have gone to such lengths in storming my dorm room over what they knew to be a nonviolent and non-drug-related search? Something, experience perhaps, tells me the situation would have been handled much differently.

The police made sure I saw that they were armed. Though they didn't pull them, they kept their hands close to their guns, terrifying all of us. I wasn't armed. Why do they have to have their hands so close to their weapons? I was just a freshman living in a university residence hall. Were they going to shoot me? Was my death going to be on the news like just another young black man's death?

The lead investigator was Officer Josh Rainey of the University of Alabama Police Department. I'd think of him later when I met Tankersley; I saw in both of them the same exuberance for arrogance and condescension; something tells me, experience perhaps, that it had to do with me being a young black man, over whom they enjoyed wielding power. Rainey seized my iPhone 6 cell phone, PlayStation 4 console, and MacBook.

"Where were you?" he asked regarding my whereabouts at the beginning of the raid.

I knew I didn't have to answer any questions, but I wasn't thinking clearly. A million thoughts ran through my head, and I wondered how events had led to this point. I was also still a little freaked out about the guns.

"I was at my girlfriend's place," I replied.

"You got any computers over there?" Rainey asked. "Yeah," I confessed.

"Take us over there."

Legally, I wasn't sure if I had to. If the warrant was just on this address, then maybe I didn't have to take them anywhere. But again, my mind was racing. Rainey informed me that I was going to be detained, so I wasn't arrested in handcuffs. Next, he threatened to take Anfernee's girlfriend to jail if she tried to log into any of my social media accounts. Rainey, though, did allow me to make a call on her phone. I called my grandmother to let her know what was going on before Rainey cut the call short.

Naively or wisely, take your pick, I led the cops to my girlfriend's room at Riverside North. Her roommate answered the door, and I tried to signal to her to hand them my girlfriend's computer rather than mine. By now, I knew who was behind the raid, and it wasn't initiated by the University police. As cocky as he was, Officer Josh Rainey didn't know he was being used and was somebody else's tool. And that toolmaker was Pearson Education.

I'd tried to avoid handing over my computer because I wasn't convinced that Pearson wouldn't get their hands on it, and I didn't believe that Pearson had any right to it. The police may have had a technical right to seize it through the warrant, but I couldn't trust the police not to hand it over to Pearson because I knew that it was Pearson that had set this whole calamity in motion. Plus, I still wasn't even sure if the warrant had given the police that right because it was in a different location. And I certainly wasn't in the mindset to accept Rainey's word for it had I asked him. Yes, I could have read the warrant, but there's another reason why the police often use shock and awe tactics: they want to grab and do everything they can before people have a chance to think clearly.

The roommate didn't pick up on my signal, and, thus, she handed over my laptop, which ran the Kali Linux OS. Kali Linux is known for its security testing platforms. I was using it for my business, so that's another reason why I didn't want Pearson to get their hands on it. What would stop them from stealing from me?

I was taken to the University's holding center where Rainey began to interrogate me.

"How did you make this site?" He was referring to a website I had created in December 2015. By now, I had my wits about me and was thinking more clearly.

"I'd like to speak to my attorney," I told him. Rainey said ok and left the room.

After having to wait a while, I reached out to Julian McPhillips, my grandparents' lawyer. While on the phone, he told me to ask what I was being charged with. Rainey replied, "Computer tampering," which is a felony in the state of Alabama. Rainey held me for four hours before releasing me to my family, who had shown up to the station. But Rainey couldn't help but get his last shot in: "You know more than you're putting on," he said. "If you change anything, I'm gonna charge you with tampering with evidence."

The University Piles On

To make matters even worse, the semester was ending. While I was dealing with my legal troubles, I had exams coming up and papers due. I also began to realize the enormity of the situation, though I wouldn't fully comprehend it until later. Somehow, I managed to finish the semester, but I couldn't wait to catch my breath to consider my next move.

After my release, Mr. McPhillips put me in contact with attorney Joel L. Sogol, and he took over my case. Within a few days, I realized what Pearson was up to.

Pearson Education had filed a complaint with the University of Alabama regarding the program I had created in December 2015. Rainey told Mr. Sogol the police had to analyze all my seized devices and equipment and that I would get these back "no earlier than two weeks from the date of questioning." In fact the UAPD did not return my property for many months. On May 23, 2016, almost a month after the raid, I received an email from Google Investigations. GI informed that the UAPD had requested access to my Gmail accounts. The request accompanied a search warrant. GI let me know that they were going to hand over my accounts to UAPD in seven to ten days. I was dumbfounded. What did my email accounts have to do with any of this?

To compound my problems, on June 8, I received an email from the University of Alabama Student Conduct Director, Todd Borst. Mr. Borst wanted to meet with me about this case. Turning to Mr. Sogol, I learned more about my mounting legal and academic problems. The program I had created utilized school resources to do its work, which the University now seemed to have a problem with. It was becoming obvious to me that Pearson Education was using its influence on the administration to get to me. I let Sogol speak with Borst to explain that my program was no different than students using Google, Photomath, MathWay, or other resources in working homework problems.

While waiting to hear further from Student Conduct, Sogol told me that Pearson Education was now threatening suit if I did not shut the site down. By now my legal team had expanded, and two of my lawyers thought Pearson was bluffing because they had little to gain but the source code I created. In the meantime, Pearson's attorney, Will Hill Tankersley, had gotten hold of my emails. How? Had the UAPD leaked them to Pearson's legal team?

In any case, Tankersley sent me a cease and desist letter, listing a variety of ultimatums, which included demands to hand over the source code with an explanation of how it worked, donate all the money I made from my program to a charity of their choice, sign a statement that I would never create such a program again, and disable my website and remove all corresponding YouTube advertisements.

Meanwhile, through my attorneys, I asked the UAPD to return my equipment. Rainey, though, said that he may have found "something" on my computer and turned over my case to the District Attorney's Office, which in turn handed it over to the grand jury for review. This meant, though, I didn't have to communicate with Pearson Education. Creepily, Tankersley said that Pearson would try to help me if they "liked what they heard from me." Their underhanded tactics continued. Later Tankersley told my mother over the phone that I was going to jail for committing a felony. How did he know that? Was he my criminal judge and jury? Did he enjoy threatening a mother that he was going to see her son go to prison? Is this an ethical practice of the legal profession? I decided against having any more communication with Pearson and Mr. Will Hill Tankersley if I could help it. At the time, I was working for Cigital, Inc., so I sought advice from their company attorney. After I explained my situation, he confirmed what I believed: that the program I created was fair use and that I had done nothing illegal.

I returned to school in the fall of 2016. The University had not yet resolved my case, so I was worried about my academic standing and whether I would be expelled from school. Dean Luoheng Han, a professor of geography at the University and an associate dean of the College of Arts & Sciences handled UA's investigation. This confused me because computer science fell under the auspices of the College of Engineering, not A&S. I figured that Pearson Education and

Tankersley lay behind this maneuver. Later I determined, though, that Dean Han was involved because mathematics is under A&S, and UA's lawyers had filtered the case down to him. On August 26, 2016, another of my attorneys, David Schoen, and I met with an uneasy Dean Han.

"I'm just a mediator," Dr. Han announced. "I don't have any experience in this area."

Neither I nor Mr. Schoen could make heads or tails of his behavior or why he had told us this. Who had convinced him to take this meeting and in this capacity in which he confessed he had no experience?

After approximately fifteen minutes, his demeanor changed. "You're a cheater," he accused.
"You were cheating."

Attorney Schoen stood up and pointed at Dean Han and snapped, "You're being totally unprofessional." Schoen was also taking notes on the meeting.

"What are you doing?" Han asked.

"I'm writing down what you said," Schoen replied.

"You can't do that. Stop doing that," he said, then he turned to me. "You are conducting mischievous activities with the school's network." Then he made a dubious claim that he had a "tech team" look at my program, and the team explained to him what it does. He said nothing I could say that was going to change his mind.

"That's impossible," I countered. His team probably consisted of students, so I called his bluff. "Pearson Education's developers could not even figure out what I did. You're going to tell me that your 'team' reverse-engineered Pearson's and my programming? Do you realize how long that would take?" Answer: billions of years.

Dean Han then changed the topic, asserting that he had scheduled this meeting because I had removed myself from the grade book.

"Why would I do that during finals?" I asked. I then showed him emails in which Pearson Education stated that they had removed me from every system that it had, effectively removing me from the gradebook itself. I explained how my program worked and that it didn't violate the student code of conduct. Feeling the four months of attacks on my shoulders, I no longer felt inhibited about revealing what I believe to have played a not-so-insignificant role in this saga. "I hope this isn't a race issue, Dr. Han. What's happened to me isn't right." I continued. "Women are getting assaulted and harassed on campus, students are getting robbed, but the UAPD isn't sending twelve cops barreling through doors to handle those. But the moment that I or someone like me creates something, all hell breaks loose." I wasn't done. "I'm using some of the revenue from my program to help churches and do positive things for the communities I've lived in. I've lost friends to violence in these areas, and I'm simply trying to pay for school, run a business, and better my community." It was an emotional speech, but it was all true. I had used part of my money to help a gym stay open for kids.

Dean Han's demeanor reverted to what it had been at the start of the meeting.

"I'll have someone of higher authority look into it," he said. "I really don't have the authority to make a decision." Then what the hell was all this about? "My door's always open if you need to talk about anything, anytime," he offered. If I suspected something was up, by the end of that meeting, I knew something was up. Pearson had gotten to him. Three days later, on the 29th, Dean Han emailed me to let me know that the case was dropped, and the University would remove any hold from my record. I still had to meet with Mr. Borst at Student Conduct to clear up any general student misconduct, so I wasn't sure if I was out of hot water just yet. When we met on September 19, Borst revealed what I already had discovered.

"You see, Desmond," he began. Mr. Sogol had accompanied me, so Borst was looking at both of us. "Pearson Education contacted Dean Han, and, well, they made it sound like the sky was falling." Han had been the one who contacted UAPD. I'm still not sure why a dean of A&S thought it was in his purview to handle this, but that's up to the administration to figure out, I guess.

"Why didn't Dean Han schedule a meeting with me first?" I asked. "I'm not a terrorist. I don't pose any threat to the school or to any students. Is this about race? Or just about money? Am I just a number?" Borst began to look uncomfortable. Good. I'd been uncomfortable for five months. "Why isn't the school representing me against Pearson?"

"Look, I understand," he replied, glancing at Mr. Sogol. "I just want to make sure that you aren't violating any rules." We went over the student handbook, and Borst concluded that I had not violated any policy. He did not answer why Dean Han went straight to the police. But I learned that it's because he bought Pearson's narrative hook, line, and sinker without taking time to think it through.

Then a pile on his desk caught my eye. It was a thick stack of papers on his desk that looked like a book. "What's that?" I asked?

"Pearson had some team do a write up on you." I couldn't believe it. Borst let me peruse a few pages. Pearson had compiled a report on what seemed like everything I had ever done in my life.

"How'd they get this information?"

Borst would not say, or did not know, and he claimed he was not allowed to give me a copy of this report. He, however, let me look it over. Finally, Mr. Sogol and I were able to get him to sign an affidavit that I had not been found guilty of violating the student handbook. It looked like some kind of victory. But in fact my nightmare had just begun.

All over the software I had written: MathLab Answers.

CHAPTER 2

MathLabAnswers

I entered my freshman year at the University of Alabama in the fall of 2015. I both like and don't like looking at pictures of myself from that semester. On one hand, I see a kid who's happy and smiling, who thinks he's got the world by the tail. He's optimistic: he knows he's smart, and he thinks that that's all it takes to make it. He has big plans, he wants to make a contribution to the world, and he sees no reason why the world shouldn't eagerly accept what he's got to give. I can't blame the kid for thinking that way. It's how we should think. It's thinking like that that makes it possible to dream and pursue your dreams.

There's a part of me, though, that wishes that kid would just be a bit more cautious, a bit more jaded. I want to warn him that some people in the world are just going to see him as a threat. I've come to know a darker world where money and power can wipe the smile off that eighteen-year-old's face and put him in fear: fear of having his dream and even his freedom taken from him. I'm not that kid anymore.

But in the fall of 2015, the only thing bothering me was having to take a math class. That's where it all started.

They Want The Answers

Like most incoming freshmen, of course, I had to take a math course, and because my major was computer science, I had to take Calculus I and II, which UA codes as MATH 125 and 126. These courses, along with other mathematics courses at UA, use Pearson Education's software program called MyLab, or for math courses, MyMathLab, or as students referred to it, MathLab. MathLab is a supplemental teaching tool where math students do homework and other exercises in coordination with the rest of whatever math course they are taking. According to Pearson, the software "offers various opportunities to practice problem-solving skills, which can help students improve their scores over the length of the course." UA, of course, isn't the only university to use MathLab; in fact, according to Pearson's own website, over 11 million students use course versions of this software each year.

While using MathLab in December of 2015, I got curious about how the program worked. That's just me. I get curious about how things work, especially when it comes to computers and programming. Ever since I got hooked on computers when I was a kid, I sit up all night many nights staring at a screen, typing, and clicking away just to see how whatever it was I was looking at worked, how it was put together, and what made it function. And college accommodated that habit. I'd be trying to do my homework or study, but my mind would wander, and I'd be more focused on the interworking of a website than I would my studies.

Plus, I was still trying to figure out how I was going to make money, and I'd get dreamily carried away thinking about how to put my skills to work.

So that's how I looked at MathLab: I wanted to figure out how the site worked, how it functioned, and how it was created. I wanted to know how the programmers at Pearson had pieced the program together. If this was a product that everyone had to use, I wanted to know what made it so usable. Plus, I wanted to think not just about Pearson's site but also about sites like Google that students used to help them with their courses. If I could figure both out, perhaps I could think of ways to combine different elements into a creation of my own. I wasn't looking for anything proprietary, but just the basic mechanisms of programming that are common to just about any program.

Then the question dawned on me: What does just about every college student who takes introductory math classes want?

They want the answers.

It's no secret that students use Google or other programs all the time to get answers to homework, especially to math courses. There was a demand here; could it be an opportunity for me? I had to find out.

I Googled Pearson Math to see if Pearson had any other websites—and it had plenty. Most of them had the same layout as far as logging in. With Pearson Math, "pearsonmath" was the username and password used for demo sites, thus making the sites publicly available. Pearson advertises the MyLab product to people who homeschool as well as schools in general, and anyone who's interested in it can access it. Basically, the sites say "Hey, if you want to see what our software looks like, if you're interested in buying, use this demo site, get familiar with it, and, then, you can purchase."

Once I logged into the Pearson Math demo account, it gave me an option to take one of three roles. I could be an administrator, an instructor, or I could be a student. The student tab looked like the one I had for my actual class that I was taking at UA. I looked around the administrator tab, but it didn't interest me. It was the instructor tab, though, that intrigued me because you could create a class and look at a grade book.

Looking into the grade book I saw in the URL an ID number that pointed to a class ID. When I incremented that number by one, I was taken to another class. If the URL ended in 97, by changing it to 98 or 96, I could see another class. And I could see demo data and production data (or prod data). Prod data are data that are available to the public and customers, perhaps around the world, whereas demo data are data only used for employees of the company. Because the demo data were not distinctly separated from the prod data, by adding one to the ID number, I could access a live class. The actual class sites were publicly accessible, so I didn't even have to be logged in to any of the classes to view them. This meant that a Google crawler—a crawler is a tool that goes to websites and pulls information—could also access this information and save it as cached data.

Using a web browser extension called tamper-data that lets you to intercept and edit HTTP/HTTPS requests and responses as they happen, I discovered that my class ID was being sent out every time that I went to a new question in MathLab. So, I took my class ID, put it in the URL, and saw my grades, my friends' grades, and my class's grades. By using that increment-by-one trick, I was able to see the same for other classes, the grades, what students made on tests, and everything that the instructor of the course could see.

All students' names, grades, emails, and student ID numbers were in plain sight.

And it wasn't difficult to discover this. Almost every company in the world sets up their data that way. It's like stacking boxes together; if you add or subtract a box, you get something a little different than before, but it's all part of the same structure. Thus, it's not uncommon to have demo data and prod data connected to the same sequence used to build both. By observation and common sense based on my experience with computers, I could see how Pearson's sites were working.

But I didn't want to look at my peers' grades; coming across that data was inadvertent. To really understand the Pearson software, I wanted to see if I could access the answers to the questions. I focused on the Flash player that Pearson had written. Suddenly, I realized that I might be able to write a program that could find the answers.

Finding the Answers

I set about to reverse-engineer the player with an FFDec, which is a decompiler tool that effectively translates a program into new source code. I did this so I would be able to understand what I needed to do to write my program. In the player, a value called privileged mode had a hook that allowed me to make a JavaScript call to change a value in the flash player. Once I figured that out, I told my Java program to change that value to one. When the value was loaded to one, I then got access to the developer tools. I then saw a point that you could call either "check your answer" or hit "show answer," and I wanted to see if I could make that call. I discovered that through JavaScript, I could make that call, or see the answers, but without losing points, as a student would if they were normally using the program by submitting the wrong answer to see the correct one.

More interesting to me was that when I right-clicked, I saw an option for "show TDX info." When I clicked that, a developer menu came up with Pearson's internal information that I didn't fully understand. So, I Googled "TDX Pearson" and learned that it was an XML format. Returning to the site, I discovered in the TDX menu an option for "current inputs," so I clicked that and that showed me the answers to put into a box. For example, the associated questions would be displayed as SA1, SA2, SA3, and so on. Then it would give you the answer to SA1 or whichever as "B" or if it's fill-in-the-blank it would tell you what to fill in. If it's a graphing problem, it tells you how to configure the graph, such as putting a line at XY and putting another line at Y2 or whatever the correct answer was for that question.

Now I had a program in place that could access answers to Pearson's MathLab. I let a few friends in on my little secret, and they took advantage of what I had found. My secret, though, started getting around, and I was getting inundated with requests from students to help them get the answers to the assignments. But I already saw financial opportunity in what I had discovered, and I was already trading my knowledge for a few forms of payment. For example, I helped some of UA's football players in exchange for Muscle Milk for working out. That, though, was small time stuff, and I already had bigger I knew that at least a million students a year, if not more, were using Pearson's MathLab.

That's when I created MathLabAnswers. I created it on Christmas Day 2015. MathLabAnswers was a Java program that solved any problem by Pearson's MyLabPlus software. The program was like other applications such as MathWay, Photomath, and even the Google search engine. But the code I created was original, and I saw it as a fully legal business competitor to other programs and options that students already had. I had not hacked Pearson's website, nor had I used fraudulent or illegal means to obtain the information I uncovered. All the sites I visited in my exploration of Pearson's products and functions were publicly available and accessible; I didn't need to steal passwords or use any nefarious means to gain access to areas of Pearson's programs. I just had to follow the path that was left for anyone to find.

But I had to address a fatal flaw in my plans.

In January 2016, Pearson started using a new HTML 5 player. I had noticed that they were setting up for its debut because they announced it on its website. Because I had planned to create my MathLab program, this became my next project. There were two versions of Pearson players: one I had figured out and one that I now set out to. In the new player, I saw a website called intellipro.com. I didn't know if Pearson owned it or was just a client. It seemed like it was the company that programmed the new player because whenever I did my homework, I noticed an IFrame, a web page inside of a web page. I right clicked to view the source and noticed that it sent me to a different page, and this page was in the Pearson domain. I took the header of this website and put it into Google, and this web page came up called player.intellipro.com. The ending of its URL had the same query or ending as the Pearson homework page—or the IFrame page on the Pearson site. I concluded that these were basically the same site. Then, I Googled the player.intellipro page and started to see other pages, like player test, player demo. Then another Pearson CMG site came up. I realized that these were all developer pages for the new HTML5 player.

When I dug further into the IntelliPro site, I discovered a blank web page. I didn't see math questions from which to retrieve answers. Usually when you go to a page like that, you'd see test questions. So, I tried to figure out why no questions were on this page while questions were on my homework page—again, because they were basically the same site, I was confused. That's when I used Tamper-Data and went back to my homework, refreshed the page, and I watched how the data was being communicated between their server and my computer.

I noticed a sequence node, a node that seemed like a bunch of random letters and numbers. But I knew it wasn't random. The sequence node, I realized, was how they loaded in a particular question for a particular session, so two students would not have the same sequence node. I wanted to know how to make a sequence node call, which led me to review some of the JavaScript calls. You can do this with "inspect element," right clicking on a website and hitting "Inspect," and then going to a console to look at stack traces, a kind of report of activity. The stack traces show what calls a program made. For example, if you were using a calculator and pressed the number one, the stack trace would say something like "button 1 has been pressed."

Next, I went to their page and saw a JavaScript call named "load sequence node." I copied my sequence node, went to the player.intellipro page, and made the JavaScript call, pasting my node in. A question came up. But I noticed that the answers were different. If the question on my student page asked, "How many foxes are there?" my page would have the answer as "Three." For the same question on the player page, however, the answer would be "Six." I couldn't account for this discrepancy. But when I went back, because I left the question, it saved my state. Now, when I reloaded the same sequence node, the answers matched. This meant that I had to load my question page first before the other page to get the answers to match. In other words, the IntelliPro site needed me to make the first move to reveal the answer to my question. That's why the page, at first, was blank, and that's why when I pulled up the answers, they were different. But if I loaded my question first, the system would know that I've been on that question.

All this took me about a month to figure out. But I now had what I needed to improve my program and keep up with Pearson's updates.

Financially, it was crucial for me to learn how the new HTML5 player worked because I had only made about ten dollars so far. I had my website, MathLabAnswers.com, set up so students could begin using it, but not much traffic was coming my way.

Not only did I have to update my site, I had to figure out how to make it pay off. I decided that my asking price of nine dollars an assignment may have discouraged college students, so I decided to reduce the price to three dollars per assignment. So, how it would work, if a student needed answers for an assignment, they could log on to my site. I had set it up so that they could only obtain the answers for one assignment, so if they wanted answers to another assignment, they would have to pay for another session. I also allowed students to buy assignments in bulk at a discount.

After reducing the price, I still didn't see enough revenue, so I concluded that the problem was lack of advertising. Nobody was going to pay for my service if they didn't know about it. I understood the benefits of my program, so I knew who would benefit from it—and I knew where to find them.

Finding the Market

I wrote a bash program in Linux using Wget. Wget goes to a web page and downloads the source of that web page. I wrote a loop for it that went from one to four million. The Wget grabbed every class in Pearson's system and saved it, and I took that data and wrote another program that went through and found any student who had any overall grade less than a B, placed their emails on one text file, and put them on an email list: first name, last name, email. Using a SendinBlue account (SendinBlue allows you to send mass emails), I then emailed everybody on that list, which was approximately 800,000 students worldwide who were using Pearson's MyLab classes. The email stated, "If you need help with your Math Lab classes, check out our website," signed MathLabAnswers Support.

I never disclosed a student's grade, so the recipient of the email did not know why they were selected for the MathLabAnswers email.

Eventually, I built my client list to about six hundred students. But in the spring of 2016, I noticed that Pearson had made a new update to their program that rendered my site obsolete. I needed to update my program again. But that's when the UAPD raided my dorm.

At this point you may be asking, but wasn't the raid to be expected? Hadn't Desmond committed a crime of some sort with all this hacking of Pearson's code?

No, I hadn't. Clever code detective work is not in itself illegal. I had, in fact, scrupulously avoided doing anything remotely illegal. I was convinced at the time, as was ultimately confirmed, that I had built a perfectly legitimate program using only publicly available information and absolutely legal techniques in navigating Pearson's sites. I used nothing that wasn't a feature Pearson provided to its customer.

But wait: hadn't I cold-emailed 800,000 people using a list I had not been granted access to?

Yes, there I will admit to being an aggressive marketer. But Pearson had effectively made the list public by failing to make any serious effort to protect it. I took advantage of that, and again there was nothing illegal in my doing so.

So why did Pearson go after me and send the police to raid my dorm room? It occurred to me that I had exposed security issues with Pearson's website that compromised confidential information of which it was a steward: Pearson was making students' private information publicly accessible. Although I was able to access student data, I never put the privacy of that data at risk. Pearson did, and that wouldn't make them look good. I concluded that they were targeting me to shut me up about their systems' laxities.

In any case, I wasn't going to be cowed by their heavy-handed tactics. I went back to work on my program.

Because the police confiscated my laptop, I had to start over from scratch to bring MathLabAnswers up to date. I went back to the IntelliPro site where I noticed that they had practice questions. I right-clicked on one of the questions and viewed "source' or "inspect element console." There were variables called "player settings" and "player state." I recall that the player setting was a BASE64 encoded string. I knew that if I took the BASE64 encoded string, decoded it into a byte array, and finally converted that byte array into a string, it would be readable text. In the player settings I looked for the extension of a file, IPX, TDX, or NTDX, which looked like a URL that didn't have a WWW in front or HTTPS. I also saw a set of numbers whose lengths did not change. So, I added a forward slashes where deemed necessary to convert the malformed URL into a usable one, making the program go to the internet and fetch an IPX file (like a zip file), which was a file they used that was publicly available. When I unzipped the file, I got a bunch of TDX files that were like XML spreadsheets. I was basically seeing a schema for the questions. Something in brackets, variable, then the name, like A. Next, text like, "the fox walked to" wherever—and then another variable. The variables had a tag name and attributes. This helped me understand the structure of the questions, but I didn't get the answers from that. For example, in one of the setting's files, it said "variable with the tilde mark." IntelliPro uses tildes to mark off variables and expressions, so it'd be tilde X1 divided by tilde X3, but the values are in that schema.

The player state was also a BASE64 encoded string. That's an industry standard for transferring data over the Internet without losing information because different systems, like Linux and Windows, interpret the same byte array differently, or the byte array means something different from system to system. Whenever I took the player state Base64 encoded string and decoded it, I got another byte array. To get the initial variables and values, I used a HashMap, which mapped names to values. The player state had the same variable names as the player settings, and it had the values right below it. I compared the variable names in the player settings to the names in the player state. This means that every time a question is loaded you get new player settings and a new player state, and that means that that question's state matches the setting which matches the question that's on the screen. From there, I had to build a pattern that matched theirs. To find the initial variables, I saw that the hexadecimal pattern Pearson used for variable names was -1, F, 15, a hexadecimal value, and then 5. Since the player state had so many instances of this pattern, I placed the byte array into a hexadecimal editor called Okteta. Okteta had the functionality to highlight patterns. So, it became very easy to identify every time a variable name appeared. For example, if it showed X3, there are always five bytes in front of it. Then, the value after that is the length of the variable name, meaning if the variable name is three letters, then the length would be three and so forth. I applied this same arithmetic to obtain the variable values, as the hexadecimal pattern that Pearson used for variable values was quite similar.

The pattern for the variable values was 17, 1, 2, 1, 5, and it also had a length and an identifier that said this value matches this variable. I concluded that if X3 is here and the first value is seven, I know X3 is equal to seven. I made a program that printed this out. Now what I had to do was to take those patterns and build a program that contained a parser, a tokenizer, and calculator or evaluator that read Pearson's XML format. A parser can take an expression and solve it without you manually typing it. The parser then went through and found the expression. So, for Question 1, if it said, X1 divided by X3, it'll first grab the values of X1 and X3 from the player state and store it, then would then tokenize the expression X1 divided by X3 found in the XML spreadsheet from the player settings with the values of X1 and X3 from the player state, and finally evaluate that expression, resulting in the answer. After realizing the pattern and mechanism, I knew I had figured out Pearson's new HTML5 player and was ready to launch the update to my MathLabAnswers.

That's when I got a cease and desist letter from Pearson Education's attorney, Mr. Will Hill Tankersley from Balch & Bingham, LLP. The date was June 14, 2016. I was eighteen years old.

CHAPTER 3

Cease and Desist

When you start a new business, you have to be prepared for the reactions from other, potentially rival, businesses. This is especially critical if these rivals are large corporations and your business entails a relationship to what their business is doing. But I didn't see Pearson Education as a rival. MathLabAnswers needed Pearson to exist: it needed students using the software and working their math problems for there to be customers to my site, which provided answers to the problems posed by Pearson's MathLab. So my goal was never to take any business away from Pearson. My business was created to give students options about how to complete their assignments.

If I saw myself as competing with anyone, it was with other programs, like Google and Photomath, that allow students taking online tests, quizzes, or completing assignments to get answers to math problems. With PhotoMath, for example, you take a picture of a math problem and it solves the problem, providing you with the answer. But PhotoMath can't solve word problems. Other programs, like Microsoft Note and Socratic, can be used as well, and something as simple as Googling the problem can also get you the answer.

What made my program unique was that it was specifically tailored to students using Pearson Education's MathLab. Students using my program knew that they were getting the correct answer and getting it quickly. And it provided the answer to word problems, too. Although it had competitors, my program was unique in that it solved all problems from Pearson's MyLab program.

One thing my program did not do was to guarantee that students would get a specific grade in their math class. And students knew that just by giving them the answer my program was not going to help them learn how to find that answer. If, when presented with the result, they decided to work out the problem for themselves to see if they could reach that same result, my program would really help them succeed. But the choice for learning the logic and principles of formulas or problems was entirely up to the student-just as it was for students using the other programs.

These other programs already existed, apparently without any trouble from Pearson. Pearson was not trying to shut down Google, of course, nor was it, to my knowledge, hassling PhotoMath, Microsoft Note, or Socratic. And I'm sure no one from those companies were ever raided and detained for their similar programs.

So why did Pearson go after me?

What Were They After?

I concluded that Pearson must have been shocked at how well my program complemented theirs. If so, they would have wanted to know how my program worked, and whether it was doing harm to their software. They would have wanted to know if I had hacked into their system. And those would be reasonable concerns.

But Pearson's reaction was something I could not fathom. Because none of the areas I accessed with Pearson's implicit permission through granted and given passwords were protected or prohibited by Pearson Education. The site never had a disclaimer, for example, that indicated that these were secure or unauthorized areas. In fact, the demo sites were invitations to the public to explore their sites. Furthermore, the demo sites were there specifically so the public could explore how their sites work. And wasn't that exactly what I was doing?

Granted, nobody but me, probably, ever explored their sites to the extent that I did. That required a certain level of knowledge and skills, in addition to a high degree of curiosity about how their sites functioned on such a detailed level. But anyone with the requisite skill and curiosity could freely explore and access these sites. Pearson had to know this.

I want to make this clear: I never harmed, corrupted, or changed the sites while I was visiting them. I observed their functions and obtained information from them in terms of their basic operating logic. I didn't have to crack or circumvent encryption, nor were there any username or passwords that I spoofed or stole.

In fact, in January 2016, I even sent the lead developer, or program manager, at Pearson, Mark DeMichele, an email requesting information about the second version of the player; I wanted to understand how it worked. I saw myself sending the email as one professional requesting information from another. His only response was "Who is this?" and I can see that it wasn't in his best interest, employment-wise, to answer my question. But my point is that I clearly wasn't attempting to deceive Mr. DeMichele or nefariously obtain secrets.

So I was legitimately surprised when I got hit with a cease and desist letter demanding that I disable my website and turn over my source code.

Based on legal advice from my attorney, Mr. Joel Sogol, I decided to comply with the cease and desist letter and shut MathLabAnswers down pending the legal outcome of my case, though I admitted no wrongdoing in doing so.

Mr. Tankersley's cease and desist letter made more demands, while referring to my business as a "Cheating Business." It demanded that I reply by June 21, 2016. It accused me of a litany of crimes, both state and federal. But also buried in the letter was stipulation #2 of what they considered compliance: "Cooperation with Pearson to understand fully what Mr. Jackson has done and how he did (and is doing) it. This would include, among other things: full access to any code or written material, associated with Mr. Jackson's Cheating Business..." These lines caught my attention because Pearson's own software engineers clearly could not figure out my methods. They did not understand what I was doing. Pearson wanted to use the law to force me to "teach" their "experts" how to navigate through their own websites.

But if their own engineers didn't know what I was doing, Pearson and Mr. Tankersley didn't have any idea if any law had been broken.

The Letter

It was quite a letter: it seemed to me, at least, disingenuous, unethical if not illegal, threatening, unprofessional, and just poorly written.

Disingenuous? Mr. Tankersley wrote, "Our understanding is that Mr. Jackson's actions have attracted the (unfavorable) attention of law enforcement. Also, our understanding is Mr. Jackson may be facing expulsion from the University of Alabama." But Pearson was behind my April detainment, as later confirmed in my September meeting with Mr. Borst of the Student Conduct office. The letter tried to make it seem as if law enforcement had acted independently from Pearson, which it had not.

Unethical? Regarding my potential expulsion, I wondered how Pearson obtained the academic or conduct information regarding an enrolled student at the University without that student's permission. My attorney, Mr. Sogol had spoken with Mr. Borst in June, and to my knowledge I had not been threatened with expulsion, but even if expulsion had been brought up during this conversation, how is that Mr. Tankersley's or Pearson's business? What right did they have to make such disclosure about my academic or conduct standing at UA at this point? Pearson hadn't filed a lawsuit. I was wondering if my right to privacy was violated. I thought of this later during my September 2016 meeting with Mr. Borst and the large, booklike file, or dossier, that Pearson, and I reasonably assert that their lawyer, Tankersley, had compiled about my life.

What information regarding my childhood was exchanged with officials at the University of Alabama? I was only eighteen at the time this dossier was compiled, so this meant that Pearson and its lawyers at Balch & Bingham such as Mr. Tankersley, if Mr. Borst was being forthright, had dug into information from my childhood, when I was a minor, and shared that information with Mr. Borst and with other officials at the University. How ethical is it in the legal profession for a private law firm that has not filed a civil suit, which Pearson had not and never did, to root around in an eighteen-year-old's childhood and then divulge that information via an extensive dossier to another party without that person's consent and with the apparent goal of using that information to persuade the third party (in this case UA) to contact the police and begin a criminal investigation.

And the dossier on me, having been given to UA, could have also served as a way for Pearson Education to influence UA administrators' decision on my expulsion or using the possibility of expulsion as leverage to compel me to comply with Pearson's demands. Then, I also recalled Dean Han's behavior during our meeting in August. After those first ten minutes or so, he had jumped quickly into accusing me of cheating—had he been aware of what was in the dossier? It bears repeating, though, that if this was Pearson's ruse, it did not work, because UA cleared me of academic and student misconduct.

Unprofessional? The letter reminded me of Pearson's earlier clumsiness of getting Dean Han from the College of Arts and Sciences involved. First, Dean Han was the associate Dean of A&S, so he was not the ranking Dean, which was Dean Robert Olin at the time. Second, Dr. Han's expertise was in geography, nothing computer-related. So, why would Pearson enlist the help of someone who didn't know have computer expertise? Why not go to the College of Engineering and ask an expert from that department? This was either a mistake or intentional. If it was a mistake, then it was a rather silly and careless one. If it was intentional, it tells me that Pearson did not want to get real computer science or engineering experts involved in the case because they may not have liked what they would have heard. Either way, Pearson and its lawyer made a poor tactical decision because Dr. Han had, after his initial aggressiveness, backed down, declared his ignorance, and he and Mr. Borst determined that I had committed no academic wrongdoing.

The letter correctly identifies me as Mr. Jackson; however, later, when levying a moral and ethical screed against my character, Mr. Tankersley refers to me as "Mr. Johnson" (as in "Mr. Johnson's actions are an insult to. . .,"etc.). I suppose to some people this was a simple mistake, but this was not a long letter. How difficult is it to get my name right before sending out an official legal letter to me and my legal representative or consultant? This gave me reason to further question the professionalism of the law firm of Balch & Bingham, and it would not be the last time I would question it. If you are going to levy ethical and moral attacks against people, shouldn't you at least take the time to get their names correct? I am not, nor have I ever been, Mr. Johnson. I am not some interchangeable human being that can be Mr. Johnson just because a lawyer is too lazy or thinks it's not a big deal. Well, my individuality and my identity may not be a big deal to Mr. Tankersley and Balch, but it is a big deal to me.

Threatening? It was during this time that Mr. Tankersley told my mother over the phone that I was going to jail, which of course gave me little reason to respect the law firm or the client he represented. Finally, as mentioned earlier, the letter referred to my business as a "Cheating Business," meaning that Pearson and its attorneys were already concluding that I and other students were cheating. At this point the University had made no official determination that any cheating took place (and later cleared me of any such claim). The "cheating" adjective, thus, was falsely used in the letter to increase its threatening tone.

Confronted with such tactics, rhetoric, and careless errors from Balch & Bingham, I considered that this firm might be one of these shady firms that you hear about, or more specifically, I thought that Balch was acting like a shakedown firm. Shakedown lawyers use aggressive and highly unethical and legally questionable tactics to scare and intimidate people into compliance. If this was what Balch & Bingham was doing, I wasn't sure how seriously I should take their communiqué.

Of course, I knew that Pearson Education had the money to hire a powerful law firm with extensive resources. Though I considered Mr. Tankersley and others at Balch to be behaving like shakedown attorneys, I also understood that I could not laugh them off—they meant to take me down, and it didn't matter if I was Mr. Johnson or Mr. Jackson or if they had to terrify my mother in the process.

CHAPTER 4

The Charges

After receiving the cease and desist letter from Mr. Tankersley and with criminal charges against me pending, I began to fight back. I consulted with attorney Joel Sogol from Tuscaloosa, AL, who had received a copy of the C&D letter from Balch & Bingham. In addition to Mr. Sogol, I still had attorney David Schoen, who responded in kind to Mr. Tankersley. Schoen knows my grandfather and is a nationally known civil rights and civil litigation attorney. He was suspicious of Balch & Bingham's tactics from the start. Schoen sent Tankersley an email on July 20, 2016 and expressed my willingness to sit with Pearson's representatives and discuss my program, the nature of MathLabAnswers, and the services it offered in a way (I'm quoting from the email) "that would be both helpful and useful to Pearson." But we made it clear to Mr. Tankersley that I would not engage in such discussion as long as I was facing criminal charges and (at the time of Schoen's communiqué) in danger of facing disciplinary action by the University, both of which had in fact been initiated and encouraged by Pearson. The criminal and disciplinary danger, Schoen demanded, had to be off the table.

Attempted Negotiation

We offered to meet with Pearson and Tankersley whenever it was convenient after August 1, 2016—after I had completed my summer job and returned to the University. We made several demands of our own. First, we demanded a non-prosecution letter, protecting me from criminal charges and that I should have transactional immunity, which would allow me to speak and discuss freely with Pearson about the situation without danger of a repeat of or more criminal charges. Next, we demanded that Pearson explain to University officials that I did not engage in cheating or academic misconduct and to the extent of Pearson's knowledge that I had "not committed any infraction that should trigger the disciplinary process." Finally, we demanded that all my items seized should be returned to me "immediately."

Furthermore, Mr. Schoen made it clear in no uncertain terms that it was Pearson that had orchestrated the criminal and academic investigations "based on information provided to them [the police and the University] by Pearson and at Pearson's impetus." Schoen cautioned Pearson and Mr. Tankersley that "it would be a terrible mistake and terribly wrong for Pearson to pursue action against Desmond in either of these spheres [criminal and academic] or in civil litigation" and that there "would not seem to be any advantage to Pearson to continuing those processes."

With the force of Schoen's response, I felt better and more confident because I could see that my attorneys were willing to fight for me, not that I had doubted them, but it invigorated me to know that I had a strong team on my side.

But Tankersley and Pearson decided to double-down; they were not conciliatory or willing to reach an agreement unless they could call the shots. Tankersley fired back an email to Schoen and my criminal attorney Joel Sogol demonstrating their entrenched position. Tankersley wrote that Pearson took Schoen's email "to be a refusal" to provide me for an "on-the-record unsworn interview" unless Pearson could assure no criminal prosecution and explain to UA that I had committed no academic or student misconduct to its knowledge. Pearson also wanted me to give them an unsworn interview under a Kastigar letter from the prosecutor (a Kastigar letter is a type of proffer letter that explains the conditions under which a defendant supplies testimony). Then, if Pearson and its attorneys were satisfied with my interview, then and only then would they ask the prosecutor to go easy on me. Second, Pearson would also enter negotiations after the interview to discuss avoiding civil litigation. Tankersley's email then stated that it was my behavior that had "caused this situation." Tankersley and Pearson also had the audacity to claim that Pearson had no role in neither my troubles with the police nor my academic problems at the University and that such an insinuation was "offensive."

This last claim angered me because it just reinforced for me that I was not dealing with honest people; this law firm was still acting like a bunch of shakedown attorneys. I was told directly by officials at the University of Alabama that Pearson had engineered this whole fiasco, which Schoen alluded to in his email. If you recall, Dean Han had told Mr. Borst that Pearson, in its initial communications with

UA about this situation, had made it sound "like the sky was falling." I was then told by an employee at UA that a representative from one of Pearson's main US offices had flown in, presumably from New York, to meet with Dr. Han directly and to express the severity of the situation, prompting Han's call to the University police. Plus, Pearson and its lawyers had provided UA with the dossier on me that, I was told, gave probable cause for the raid. Pearson had so startled UA with its aggressive and outraged response that UA attorneys had to review the case with the information that Pearson provided—all this was going on behind the scenes before I even knew what happened.

My attorneys, particularly Schoen, were taken aback by Pearson's unreasonable response; they were also surprised at Pearson's team's denial that it had any role in the criminal and academic troubles I was facing. To be frank, that denial was utterly shocking in face of the evidence that was provided by the University and the UAPD to the contrary. My attorneys were surprised at Pearson's aggression toward me and outright denials of plain facts. Even in their private discussions, my attorneys felt that Pearson and Tankersley were trying to bully me into total submission.

With my legal team's response, the criminal and civil situation then became a waiting game. Periodically, Mr. Sogol, who was handling my criminal charges, queried the UAPD and Officer Josh Rainey about the pending charges. Officer Rainey told Mr. Sogol that the investigation should determine quickly my case's disposition; however, "quickly" turned into weeks, which turned into months, which turned, academically, into semesters. It is said that the wheels of justice grind slowly, and indeed, a year, especially for an eighteen-turned nineteen-year-old college student, it seemed that the wheels ground to a halt.

I faced an uneasy dilemma about what to hope for: a quick end to the investigation could lead to charges, but as long as the investigation was open, it meant I was free, even if in limbo. Of course, a swift end and exoneration was optimal. At the end of September 2016, I was at my most hopeful because I was cleared of academic and student misconduct by UA, but as the situation lingered and these charges hovered, I quickly began to sink into a state of resignation and dread. During the academic year of 2016-17, my grades began to suffer. I had done well my freshman year; I even earned an A in Dr. Love's English course. In fact, my group project in English was voted the best in the class, and my group competed in the university-wide group project presentation that spring. The situation, however, with Pearson sent my grades into a downward spiral, even though my defense as of the fall of 2016 had earned some early victories and showed Pearson that I was ready to fight back with all I had. But that's the point: it was taking all I had, emotionally and financially.

You see, I had started MathLabAnswers partly because my mom wanted me to get a job and pay for my school. My dad (technically, my stepfather, but I consider him my dad) was in the Army, and I came from a long line of people who worked hard and believed in making your own way. So, my family's work ethic was instilled in me, and I had that innate drive to be industrious and make sure that I was taking care of myself financially. I didn't, though, want to work for nine dollars an hour somewhere if I didn't have to; if you crunch the numbers, it's just not worth working for such, when trying to pay for your living and for college if you have other options. I'm not knocking students who work those jobs if that's what they must do, and perhaps, I might've found myself having to take just such a job if I had to, but before anyone should resort to such, they should consider their options. And I knew I had it in me and that I had the skills to make more money if I could put my mind to it, and that's what led to my creation of MathLabAnswers.

The Good News

Fortunately, I had some good news toward the end of 2016. I had the opportunity to apply for an internship at Google, and I approached Dr. Love, my English teacher from the spring semester, to request a recommendation letter. He gladly accepted my request and that December I learned that I had an offer from Google to work for them in the summer. Thus, despite my looming circumstances, I had something to celebrate and look forward to.

Nevertheless, all this took a toll on my family. My parents were living with my siblings in Killeen, TX at the time, so they were learning everything from afar. In the summer of 2016, my mom had a conversation on the phone with Mr. Tankersley during which he told her I was going to go to prison. My mother was worried, and she kept asking me if I had done what Pearson had accused me of. At times, my situation wore on her nerves and could make her frantic. She told me that even if I didn't think I did anything wrong that I may have broken the law without knowing it. She had a legitimate point, but I don't think she was aware how much I knew about cyber laws; by this time, I'd worked remotely for several cyber companies, and I was more versed in the law than anyone, even my lawyers, were aware. My confidence in my innocence remained strong, not just from an emotional standpoint, but from solid legal grounds. My mom, though, was justifiably scared for me, especially after Tankersley had threatened her over the phone about me going to prison. For a while, we were able to keep my problems from my grandmother, but Tankersley had also sent an email with my grandmother copied on it. It caused some of my relatives to question my innocence, so Pearson's rhetoric had divided the family. So, by now most of my family was aware of what was going on. Moreover, my uncle, though supportive, tried to prepare me mentally for prison. "You gotta steel yourself for doing time; otherwise, you won't be able to handle it," he'd say. It also cast a shadow on my dad's military reputation.

Being in the Army, he had to wonder if my case might affect his standing with his superiors and how it might look to his subordinates if they were to learn about my case.

My grandparents, though, live in Alabama, and I'm close to them. I made frequent trips to their home in White Hall, which is about an hour-and-a-half drive from Tuscaloosa. In fact, I was at my grandparents' house during 2016 spring break making improvements to MathLabAnswers. Unfortunately, later that year my grandmother had a stroke, and I couldn't help but dread whether my situation with Pearson and my pending charges contributed to her condition. You see, these are some of the realities that law firms like Balch & Bingham don't care about when they're trying to tighten the vice on one of their targets; they don't care about the emotional and health toll that putting as much pressure on someone can have on their targets and on their families.

Or, even more cynically, they may even factor such into their tactics. It's not unheard of that some legal strategies are built on grinding someone down, grinding their families down, until their target's will be totally broken, especially if the health of the target or the health of the target's family members are put at risk.

So, there I was—a nineteen-year-old being ground up by this legal process that was affecting my education, my mental health, and the welfare of my family. I just couldn't understand why Pearson and its legal team would not accept my offer to explain what happened; I couldn't understand why they wanted to see me prosecuted and sent to prison. Both I and my legal team had sensed from the tone of his correspondence that Mr. Tankersley was getting a kick out of persecuting me—his future behavior and words would only reinforce my suspicion.

As the year dragged on, so did my classes. I couldn't keep my mind focused on my schoolwork, and my girlfriend noticed I was becoming more withdrawn. Before this whole legal odyssey started, back in February 2016, my girlfriend had told me that she dreamed I was making a ton of money from MathLabAnswers, which at the time I wasn't. But then suddenly I wound up on the news over this whole fiasco. Her dream was prescient; nevertheless, she remained supportive of me throughout my ordeal.

Finally, the summer arrived along with my time for my internship at Google in Mountain View, CA. While working at Google, I worked for the Geo-Data (Maps) Team and assisted with Google Security. Additionally, I found vulnerabilities on their internal network and deployed a RAT (remote access Trojan [a form of malware]) that remained undetected as a security demonstration. I had also developed a program called MacNinja, which is a tracking program that Google's directors were interested in. They wanted the source code to the program, but they weren't willing to pay for it. Concurrently, I discussed the Pearson case with Google's legal team. From my description of my project, the lawyers advised me that they believed I had done nothing illegal. Although this was comforting to a degree, their opinions had no sway in what was going on 2,300 miles away in Tuscaloosa. I saw an opportunity to trade my source code for MacNinja for Google's legal assistance, but Google wasn't interested. Incidentally, it turns out later that Mr. Tankersley would take a keen interest in Google and its methods for handling data.

It had now been more than a year since the raid, and I had not heard anything about my case in a while. Adding to my frustration and wonder, I didn't even know what Pearson was planning in terms of civil litigation. I supposed that Pearson was waiting for the adjudication of my criminal case before making its next move, or at least lying in wait to see the prosecutorial pressure ratchet up before I gave up and started to comply with their demands—and begged them for help. That way, Pearson could act as the hero saving me from the worst of the charges.

The Indictment

Then, while at Google, I received a message on July 17, 2017 from my attorney, Mr. Sogol. The email was short and blunt: "I was able to learn today that you have been indicted... there is a writ of arrest for you. I need you to be careful and get back here so it can be served, we can make a bond... I will try to find out what the bond is before we turn in, but that may be hard." The world crashed on top of me. I remember getting on the plane in California and watching people settle into their seats, trying to get comfortable as they prepared to go on vacation, reunite with family, or head off on a business trip, and here I was returning to Alabama so that I could be arrested and charged with multiple crimes.

When I returned to Alabama, I went to stay with my uncle in Birmingham. I had agreed to turn myself in on Friday, July 21. I had to arrange for bail because if I didn't get in on time, I'd be spending the entire weekend in jail. When I was booked, they put me in a cell with some hulking dudes who probably weren't in there for having anything to do with computers. I was in long enough to witness one of these dudes knock a guy out because the latter wouldn't shut his mouth. I couldn't help but think that this could be my life every day for however many years if I didn't figure some way out of this nightmare. Fortunately, I was able to make bail, so my taste of jail-life only lasted a few hours.

Before my indictment, the University of Alabama Police Department completed its report on March 31, 2017. The report listed "Computer Tampering" as the charge and the title of the report was designated a "Felony Case Report." Under the computer tampering statute in Alabama, the charge can either be a misdemeanor or felony depending on the conditions of the case. Detective Rainey had marked the "Misdemeanor" box in the report, which later led to confusion about the nature of my indictment. Specifically, the UAPD asserted that I had stolen email addresses "during unauthorized access to Pearson websites." The notion that I committed a crime because of "unauthorized access" would be a key point of contention. Plus, through the report, I was able to comprehend more of what had transpired on Pearson's end that led to my arrest and the subsequent charges.

According to the police report, on April 25, 2016, Detective Rainey from UAPD had a conference call with Pearson's Vice-President and Senior Counsel John Garry, Chief Information Security Officer Rod Wallace, and another Pearson employee, Arthur Burns. They explicitly told Detective Rainey that Pearson was "willing to prosecute," which contradicted Mr. Tankersley's later claim that Pearson had not instigated the criminal charges against me. These representatives of Pearson stated that their "main concern" was that I had "personal information and may disseminate it." At 11:32 that morning, Rainey obtained the search warrant that led to the raid on my dorm room.

I also learned that UAPD dispatched an officer to locate me in one of my classes. This angered me because UAPD was going to humiliate me in front of the student body by having me escorted from class in front of my fellow students and my instructor. Luckily, I happened to not be in class that morning.

In May 2016, I had agreed to give a statement to the UAPD, and I provided Detective Rainey with a general overview of what I had done; at the time, I had hoped that my explanations would demonstrate that I had done nothing illegal and that I was willing to cooperate. The depth of the report revealed, though, the amount of time and legwork Pearson had invested in shutting down my website and pursuing prosecution. The following December, Pearson provided Rainey with witness statements that detailed its internal investigation that led to my arrest. On March 31,

2016, Pearson's Security Operations Center forwarded to Shawn Garringer, a forensic analyst for Pearson, an email advertising my website. Garringer and another Pearson employee, Camden Daily, then began a forensic investigation into the nature of my site to determine if I had made any "improper" or "illegal use of Pearson's systems or information." Garringer characterized my site as an "Unauthorized Solutions Website." He located my YouTube video in which I explained to my users how to use MathLabAnswers.com and claimed that I was making "unauthorized" use of Pearson's demo account. Garringer collected the IP addresses that had accessed the demo account and sought to pair any addresses with my marketing emails. He forwarded his findings to Camden Daily and Mark DeMichele, Software Development Manager for Pearson, for further investigation.

If you recall, I sent an email to Mr. DeMichele in January 2016, asking him about MathXL software. Garringer's email reminded DeMichele of that email and together they discovered that my email had originated from my iPhone, which had a University of Alabama IP address. Using the email address from which I sent my query to DeMichele, they began to run searches on this address. They uncovered postings that I had made using that email address from June of 2015, when I was seventeen years old, and before I began attending the University of Alabama. This posting was completely unrelated to the creation of MathLabAnswers.com and anything to do with Pearson. Nevertheless, Garringer noticed that my email address was associated with a username I had used over the years. Garringer shared this information with Daily and DeMichele, and Daily found that my username was connected to "a University of Alabama student named

'Desmond Jackson'." Subsequently, they found that I had posted questions on Stack Overflow, a knowledge sharing website for computer programmers, that pertained to some of the technology that Pearson used. Of course, there's nothing illegal about asking questions about how software works and learning different techniques of programmers, but Garringer, Daily, and DeMichele thought I was up to no good. They discovered that I had been using what they called the "Cheating Software" to complete assignments, though, as I stated before, I had received permission from my math instructor, Phyliscisia Carter, to use my site under certain conditions.

As the Software Development Manager for Pearson, Mr. DeMichele handled the investigation into how my site interacted with Pearson's Math XL software. His task, as he stated in the report, was to determine whether I had tampered with Pearson's systems by gaining unauthorized access or by exceeding authorization use. DeMichele's determination of "unauthorized" access and excessive use would prove an important point later—especially in relation to Pearson's future manipulation of the meaning of such. DeMichele concluded that I had manipulated and altered Pearson's software and servers in such a way, he states in the report, that was "disruptive and illegal." My site, however, did not damage Pearson's software nor did it interfere with its programs to carry out its normal operations.

The University of Alabama Police Department also brought in Scott Gilliam, a digital forensics analyst for the Joint Electronic Crimes Task Force at the University of Alabama. Gilliam examined my laptop, MacBook, flash drive, iPhone, and PlayStation (yes, my PlayStation!). All Mr.

Gilliam could add to this investigation was to confirm who I was, that I had created MathLabAnswers.com, revenue that I had acquired, general but not specific information about how my site worked, and some vague speculation about my actions and intent. Much of what Rainey had to go on had already been discovered and was provided by Pearson because nothing incriminating was gained from Gilliam's investigation of my devices. Gilliam could only point to "digital evidence" that my site interacted with Pearson's, which, of course, I never denied and even explained to Rainey in my May 2016 interview.

A Felony

When I returned from Google and arrived in Tuscaloosa, my attorney Joel Sogol and I were confused about what exactly I was being charged with. Rainey had marked misdemeanor on the police report, so the severity of the case against me was in question. The DA, though, had decided to charge me with a Class C felony stemming from the Computer Tampering charge. Section 13A-8-112 of Alabama state law defines Computer Tampering as acting "without authority" or exceeding authority by "accessing and altering" any "computer, computer system, or computer network" and "altering, damaging, deleting, or destroying computer programs or data," in addition to other actions committed. Such violations can be misdemeanors, but a Class C Felony is committed if "the actor's intent is to commit an unlawful act or obtain a benefit, or defraud or harm another." Although we tried to use the discrepancy in Rainey's report regarding his marking of "misdemeanor" to squash the charge, we were unsuccessful. In Alabama, a Class C felony can carry up to ten years in prison and no less than one year. But because I "accessed" Pearson's site over four million times, based on the emails I sent, the DA threatened to charge me on multiple counts, meaning I could be facing decades in prison if convicted.

Because of the confusion over Rainey's report and the language of my indictment, which charged me with just about everything in the statute, Mr. Sogol made a motion for a more definite statement. Sogol forwarded a copy of the indictment to Schoen, who remarked that it looked like the DA had just Xeroxed a copy of the entire statute and handed it to the grand jury. My legal team and I were thinking that the DA's office really wasn't sure what crime I committed and probably did not comprehend the technical aspects of what I created and how I created it. Consequently, the scope of the indictment made it seem like they were throwing a bunch of charges up against the wall to see what would stick. Even more confusing was that on August 3, 2017, Mathew Hudnall, the Deputy Director of the Center for Advanced Public Safety at the University, sent me an email stating that he met with officers from the UAPD and that my name came up "in casual conversation." The officers, according to Dr. Hudnall, said that "there wouldn't be any criminal charges" and that they spoke about me in "a VERY complimentary manner." Confused, I replied to Dr. Hudnall that I had been indicted. Dr. Hudnall replied, assuring me that UAPD "had thought Pearson didn't want to proceed and that it [the case] had been dropped" and that Pearson, UAPD, and the DA were clearly "not on the same page." He then went on to say that he thought "it would be bad though for Pearson if their flaws come out publicly during trial." By flaws, I knew he meant that my program exposed Pearson's lax security, leaving millions of students' private information publicly accessible.

It was becoming clear to others who knew the case that Pearson was overplaying its hand. Mr. Tankersley's initial emails referenced several federal laws he claimed I had broken; however, my attorneys and I knew that they did not want the federal government to know its transparent software was possibly violating student privacy laws, such as FERPA, and leaving such private information exposed for the world to see.

I had two more hearings later that year, one in October and another one in November, so my case continued to linger. Even more delays occurred as Deputy District Attorney Sarah Folse, the prosecutor on the case, went on maternity leave. As 2017 ended, I faced another year of uncertainty, and the waiting continued to take a toll on my mental health and even began to affect me physically. If I was withdrawn before, I was even more so in the next year. I'd go to school, come home, and play video games. My girlfriend thought I was suffering from depression, and she became worried about my state of mind. My grades continued to suffer. I once had a 3.9 GPA, but now it slipped into the C range. To make matters worse, I began drinking heavily. I'd never been into drugs or alcohol before, so turning to substance abuse was a dramatic step in the wrong direction for me. I thought about what my uncle said, about preparing myself for prison, and I couldn't help but think that maybe he was right.

In 2018, my lawyers asked attorney Daniel Fortune from Huie, Fernambucq, and Stewart out of Birmingham to look at my case. Mr. Fortune was a state and federal prosecutor for fifteen years and was the United States Attorney General's chief cyber prosecutor for the Northern District of Alabama along with being the Computer Hacking and Intellectual Property Coordinator. I went to meet Mr. Fortune with the complete confidence that I could demonstrate my innocence. Once he heard what I had to say, I knew that he'd be on my side, and under his leadership my defense was about to turn the tables on Pearson, UAPD, and the Tuscaloosa District Attorney's office. After all, I'd been sure of my innocence all along, and I believed that one of the chief aspects of the case that worked against me was the intricacy and complexity of computer codes, programs, and techniques that only I understood. Although Mr. Sogol and Mr. Schoen were in my corner, they lacked the technical expertise needed and were relying totally on my explanations for my defense. To be honest, I came to believe that I was locked in a conundrum. Only I could explain to a jury what I did, but I couldn't expect a jury to fully understand unless they had my level of knowledge about computers. I could serve as my own defense attorney, but then I lacked the knowledge of the law that Mr. Sogol and Mr. Schoen possessed. With Mr. Fortune on board, I'd finally have a lawyer who had just such expertise. After I learned about Fortune's credentials, I couldn't have been happier and more optimistic. It seemed as if Fortune's name carried a prophetic meaning for me.

Going into the interview, I expected Fortune to test me. I knew that attorneys might try to poke holes in their clients' cases to gauge their responses. So, I wasn't worried when Fortune started grilling me over different aspects of the case. Of course, with him being a former federal prosecutor, I was getting a baptism by fire. Despite his aggressive interrogation, I believed that I held my own and had clearly made my case to him regarding my innocence. I didn't have to keep lies straight; if you're telling the truth, then the truth remains the same no matter what the questions are or who's asking them. Plus, I kept in mind that this was just an exercise and that as long as I stuck to the truth, when it was over Fortune was going to congratulate me, tell me how impressed he was, and he'd take the case. For the first time, I could finally see the beginning of the end, and I envisioned a swift conclusion to this two-year saga. But when our interview was over, Fortune's demeanor remained stern and serious. No smile cracked his face. Instead, he looked me dead in the eye, and said, "Desmond, I think they got you."

CHAPTER 5

A Question of Character

As of September 2016, the University of Alabama had cleared me of all wrongdoing. The University concluded that I had committed no academic or student misconduct. In fact, not only had Mr. Borst and Dean Han made such determinations, I had already received permission to use my MathLabAnswers program from one of my math teachers, Ms. Carter. Because Pearson and its lawyers had attacked my character and had compiled a dossier to discredit me, I sought witnesses who could rebut these attacks. On August 22, 2016, Ms. Carter wrote a letter attesting the quality of my character. I also discussed the situation with my English teacher, Dr. Christopher Love, who also wrote a letter of support.

Hacker Past

Nevertheless, Pearson had uncovered a weapon, regardless of its relevance, that it decided to use against me. That weapon was my past. I mentioned earlier that my grandmother convinced my mom to buy me a computer when I was eight years old. One night, I was playing an online game, Stick Arena, and suddenly another player entered the game room. The player moved his character all over the place, through walls, on and off the screen, just flying everywhere, and performing feats that weren't within the rules of the game. Stunned and amazed, I begged the player, using the chat option, to tell me how he was manipulating the game and performing all these tricks. At first, he ignored me, but after a few minutes he told me to go to a website and flashed the URL on the screen. And just like that, the guy's player vanished. Being eight, of course I was fascinated and had to know how this person was able to break the rules of the game and do whatever he wanted. So, I took the bait and went to the website. Thinking that I was going to learn all these tricks, I couldn't wait to see what this website was going to show me. My anticipation swelled, and I couldn't help but smile at my computer screen as I waited for the website to reveal all its secrets.

To my horror though, all that came up was a pornographic website, and there I was, this eight-year-old kid staring in shock at these naked women all over my screen. I was terrified. I immediately thought I was going to get in trouble, and I started crying and ran to confess to my mom, telling her that I didn't know what had happened and that this stuff on my screen wasn't my doing. I just wanted to play a game. She wasn't mad, but my computer had downloaded a virus.

Despite the virus, I became determined to learn how the hacker had invaded and took over the game. I searched all over the internet, rummaging for techniques, game cheats, codes, and whatever I could find. Most of my time was invested in looking for ways to manipulate online games. I'd sit up all night surfing the Net, learning more and more about how computers and the internet worked. Over time, I discovered online forums and hacking sites where I learned tricks, techniques, and skills to manipulate online video games. I was also picking up on vocabulary and jargon; I was getting a topnotch education on programming, coding, cybersecurity, and the language of computers.

By the time I was in high school, I accomplished my goal of mastering online gaming manipulation. I'd enter games like Call of Duty and Minecraft and just, to put it bluntly, fuck with people's games. In Minecraft, for example, I'd wipe out everything a player had built just because I could. The power I had as a child operating in cyberspace infused me with a feeling of omnipotence, that I could do anything I wanted. I had limits, though, as I continued to learn, and one lesson I learned was that there is always more to learn and there is bound to be many other people out there who know more than you do. Later, though, I stopped picking on innocent people. As I delved deeper into the hacking world, I discovered hackers were stealing money from people by hacking PayPal, so I'd take money from them by creating a PayPal account and having their stolen money transferred to me.

While attending ASMS in Mobile, I got into trouble because I had a dispute with Enjin, a company that hosts gaming platforms. I was exploring the vulnerabilities of their site, so they shut my account down. I asked for my money back, which was only eight dollars, but they refused to refund me, so I hacked into their site and took my eight bucks back. Enjin contacted the authorities and the FBI became involved, and then the school, because I had not used a VPN whilst hacking, thus exposing the school's IP address, was threatening to act against me until I explained what happened. Two years later, though, I sent Enjin an email asking if they'd be interested in hiring me to test the security of their systems. Enjin agreed, so I worked for them until 2016. By the time I was sixteen years old, I had begun to leave my mischievous past behind me. Not only did I work for Enjin, I also landed a web security job for Nitrogen Sports, and later I started working for Cigital, Inc., which became Synopsys, based in Virginia.

When Pearson's employees started doing research on me, it's clear that they thought they had stumbled on something profound because I had posted on websites asking questions that pertained to certain aspects of their software. As its employees Shawn Garringer and Camden Daily found, my email addresses and usernames were associated with what they assumed to be hacking activities; Pearson's lawyers jumped to conclusions about these circumstances' relevance to MathLabAnswers. For example, they discovered I had a Bitcoin address and had posted to HackForums.net, a site that is frequented by hackers but also by people wanting to share legitimate information and knowledge. They traced my username and email address to my GitHub page where they learned my real name and that I was a student at the University of Alabama. GitHub is a site for developers to share knowledge about building software, and there I had begun to advertise myself professionally.

The information Pearson collected on me was passed to the University of Alabama and was the basis for the dossier that sat on Mr. Borst's desk the day I met with him. I had no doubt that Pearson used the dossier to persuade the University and the UAPD to take their aggressive actions against me at the onset of the case. The dossier was used to smear my character, to persuade university officials that I was some dangerous, criminal hacker who was breaking the law with impunity and wreaking illegal havoc on the University. Fortunately, their efforts as far as the University was concerned led to nothing more than an awkward, short-lived inquiry.

To what extent this dossier or my past in general influenced the criminal investigation, I can't be sure. It did, in fact, help give the UAPD probable cause to seek the search warrant. But I wasn't trying to nor have I ever tried to hide my past. After all, I was still using the same usernames and emails in these forums as I had before. Because I knew I wasn't doing anything illegal by creating my website, it just never occurred to me to create new email addresses and usernames. With my knowhow, I could have easily gone to great lengths to cover my tracks if I had wanted. Moreover, much of the information Pearson had uncovered was trivial, involved nothing illegal, and had come from before I had even turned eighteen.

The vagueness and breadth of the grand jury indictment reinforced what I had already known about my case: no one was really sure what exactly I had done, so my past was being used to tarnish my character and undergirded the suspicion that I must have been up to something illegal even if no one was sure what crime I had committed. Assistant District Attorney Sarah Folse would rely heavily on Pearson's software engineer Mark DeMichele's and Shawn Garringer's statements to the UAPD about the nature of the construction of my program. Furthermore, my attorneys and I realized what I was going to face when my case went to court. Pearson would turn over the dossier to Folse hoping to show that I had an unsavory online history. The more I thought about my situation, the more discouraged I became. I wasn't sure if I had the resolve to persevere through the hearings, the investigations, the possible trial, and the threat of drawn out civil litigation. I knew that my character would be on trial, and I had to fight to protect it. I knew I had to think long and hard about who I was and from where I came.

CHAPTER 6

Lowndes County

Somewhere beneath the loamy ultisol of south-central Alabama seeps up from the red clay stories long-buried, half-told, fully-lived irrepressible and effervescent; stories that will not die or be overlain and suffocated by thickening years. Stories passed on through remembrances, eulogies, and the building of memorials insist, by the very moral purpose of their persistent existence, that something must be gained and progressed, and that something must be learned and passed on as vital irrigation and seed. Pressing upward like geothermal gas, the stories of people long since gone and those who continue to live, stories that both share but were lowered in the ground with the deceased or that fell to and soaked in the soil like raindrops from the sifts of memories of the still-living find their way through the earth's pores and drift upward. They swirl, float, dip, and rise like pollen and dust carried by zephyrs and gusts, spreading out across the plains, ridges, and soft hills of Lowndes County. They may navigate, or they may meander aimlessly. They may attach themselves to the bark of swamp cottonwood or black willow or rest on the leaves of river birch and sugarberry. They may hide among the clusters of wisteria and sweet clover or lie like pebbles in the Johnson grass. Or they may intertwine themselves with the air and become part of what is breathed and expelled, or their particles might sprinkle the bricks and wood of churches or the granite or ancient slate of headstones. They may dangle from the tendrils of Spanish moss or rest at the tops of blackjack oak or in the bony arms of rare shellbark hickory. Looking northward toward the spindly coil of the Alabama River, you might catch the scent of these tales on the rippling skin of its waters or haunting the overgrown, translucent trails once traversed in secret by light-bearing marchers, organizers, and the persecuted fugitives from Southern vigilantism and terror cutting through pine and cedar as pointed, pin-branches pricked their ankles and shallow creeks moistened their feet.

If you don't sense them, then you pass by or through them unwittingly or drive past them with the velocity of the massive trucks that menace the modern split four-lane highway connecting Montgomery to Selma. If you're traveling this road, you will pass to the north the antebellum mansions of Lowndesboro and pass to the south the county seat of Hayneville, wherein you could find across the street from the town square and courthouse old men selling watermelon, corn, or tomatoes from the wooden bed of their rusty dust-caked truck. The watermelon might cost you eight or ten dollars depending on the size, or you might strike a deal if the men are ready to leave. Farther down, past Hayneville, and down another road you might find, at the dimming of the day, a congregation of men and boys gathering at the meat market in Mosses, with their cars packing the gravel lot or you might see the remnants of old corner stores with faded Coca-Cola signs on rotting, splintering white wood. If you stick to the highway, you might pass a state trooper zooming from Dallas County to Montgomery County or vice versa as if Lowndes were an inconvenient void placed to provide space between their citified seats. Heading westward from Lowndesboro, you'd come through White Hall and Benton, before exiting the county and into Tyler and a few miles before you cross the famous bridge into Selma where the brutal, futile, final assaults of the Old South inflicted bloody but temporary wounds on the New.

Other specks dot the county, and the largest one is the county's largest town, Fort Deposit, but other places like Mt. Willing and Gordonville hide on backroads and mask themselves with timber and kudzu. You have a better chance at seeing a deer than a person. The night sky drops quickly in Lowndes; its darkness is blinding, save for the few feet of headlight that beams from your car, but even then, in the rays, swirl swarms of insects turn the light into a dirty, speckled cloud. Overhead, on a clear summer night, the stars and planets shine like silver and gold glitter tossed onto a sheet of black paper, and the moon, depending on its phase, acts as a cool, gray midnight sun that only tempers the solar heat that will drench the day in a steamy shower of brilliant, fire-star radiance. Many have remarked on the magnificence of the skies of Alabama, but their magnificence comes from the deprivation of electric light from towns and cities and from the long stretches of uninhabited plantation, field, and forest that reflect nothing but the heavens' own darkness. Yet the beauty is real and sensible, self-evident by just a glance or glimpse upward into a part of the universe not spoiled by artificial light. Part of the beauty is through the contrast of galactic blackness with the defiant embers of a great succession of cosmic explosions that, when cooled and hardened in the aftermath, paradoxically hold the universe together. We might also call such a scene natural, pure, unspoiled, and it might connect us to something that we might sense in ourselves that possesses just such traits, and in that view of Saturn, Zosma, and Coma Berenices we see as much as who we believe we are as we see in and believe about them. We might imagine ourselves in the primitive ages of our ancestors looking up in wonder at just such a night sky and pondering our genesis and, then, thinking, now, in our present age, how little more we know now as they did then and that we might share the substance of some belief about who and what we believe ourselves to be. Or in the least, if we do not concede that, if we possess the arrogance and hubris of certainty about our progression, then we can at least share with them the knowledge that we, too, are specks, and even such one that we are likely invisible from those bodies that look down upon us.

And if you are there, somewhere in the county, outside, as the sun dives below the horizon and the brilliant blue of the day darkens, and suddenly you find yourself in a night so deep and rich that you cannot but see the hand in front of your face, you might begin to intuit that the stories of the past are called forth, enchanted by the sensation of you sitting alone or with someone else and, you alone or both of you, searching and finding that beauty in the starry obscurity. And you may not know the history, but you know that this place has it, and that something significant must have and has happened here and, you know, of course, that some people have lived here for millennia or passed through and hunted on its grounds, and maybe you know just the gist of that history: the Alibamu, the Creek, the Spanish, the French, the English, the Americans, and General Jackson; the South and slavery and cotton; and you might know of Jim Crow and Civil Rights and then you might know something, a little, about how the place where you are is now. Or maybe you know even less about this history as it pertains to Lowndes than anyone might suppose. But your senses are ignited by nightfall, and you get the feeling of some distant past as if it is no longer past but right there with you, around you, and knows that you are there. And that past wants to speak to you and you want to speak to it, and the only way you might know how to communicate is to look up to the sky and back into the ground-level lightless panorama and just breathe and will your mind to conjure images of what you think that past might have been like and what it might want to say. And above you, and maybe you realize this now, above you are the witnesses to that past, the stars that hovered over a time that is no longer, and that, in kind, the light they bring to you, to your eyes, is, in fact, the light that left their bodies years, decades, and centuries ago. You know that they have seen, that they have borne witness to the past that you can only imagine, and in their light shimmers the record of their own pasts, and together, somewhere in the air around you, the past of earth, your imagination of it, and the interminable past of the heavens coalesce.

Suddenly, in that moment, you sense something stronger now, and you might become afraid, afraid of the darkness, afraid of the awe that overcomes you as you sense your own insignificance, but those fears might be allayed by the stories, because they exist for you, they have been told for you, and there they come out of the night and mixing into all that is around: the light from the stars, your imagination, your presence, and you are no longer afraid but willing to listen. And it is now, here in Lowndes, that you can finally hear what all in the night has to tell you.

White Hall

This is the county where I am from, where I mostly grew up and was raised. If anyone has ever heard of Lowndes County who is not from Alabama, then they may have heard of it referred to as "Bloody Lowndes," a moniker derived from the Civil Rights Movement, and then they may know that it is part of a region of Alabama called the Black Belt, which can refer both to its soil and the predominance of African Americans who live in the area. The town I grew up in, White Hall, is close to 100% black. The county shares the history of much of the South: slavery, cotton, Jim Crow, and racial violence and terrorism that lasted well into the latter part of the twentieth century. From the nineteenth century and into the twentieth, steamboats churned up and down the Alabama River between Montgomery and Mobile, passing through Lowndes on its northern border. Black tenant farmers still worked the cotton plantations a hundred years after the Civil War. In Hayneville, several mills once stood: a cotton-seed oil mill, a grist mill, and a sawmill. Fort Deposit was a central shipping center along with Benton which sat along the Western Railway of Alabama. White Hall came later, formed in the 1930s as a government project that redistributed land from the White Hall plantation to black farmers. In addition to cotton, the county has produced dairy, beef cattle, corn and poultry, and it remains a sparsely populated rural area.

White Hall stretches from the Alabama River to Highway 80, and if you don't blink you might catch the sign for it. There are a couple of roads in, and one takes you past cattle pastures in which you might see cows cooling off in a pond, a marsh, a graveyard, and a few modest houses along the way deeper into the town. Farther back, lies the site of the Battle of Holy Ground, which took place during the Creek War in the early 1800s. There's a school and a gym and looping back to the highway you'll come to the two casinos that stand on its southern border. In the middle of two turn-offs, sits the Lowndes Interpretive Center. About a mile behind the center is where I grew up, and where my grandparents lived. Thus, you might imagine the stories that might be heard or might be told in and from such a place, and if you have an idea of what many of these stories are about, then you might understand more about me and who I am.

I wasn't much of your typical country kid, though. When I lived with my grandparents, my grandfather used to get on to me to mow the yard and my grandmother would supervise any yard work I did to make sure I did it right and didn't hurt myself. Once I got a computer, it was almost all I worked on and most of what I thought about. I had friends, but I don't think any of them knew that I was up all night in my room figuring out the intricate workings of the internet. It's something that you wouldn't want anyone in White Hall to know necessarily, and nobody in my school would find being a computer geek impressive. In fact, it might even invite trouble. I inhabited and lived in two worlds: the world in which I was just like every other kid in Lowndes, or in other places that I lived, and the world of my room and my computer. Lowndes and White Hall can be a rough place, so rough, that such a small place can produce an NBA player like Ben Wallace, a White Hall native and a childhood friend of my mom. Wallace was always known as one of the toughest basketball players in the NBA and won a championship with the gritty 2004 Detroit Pistons against Shaquille O'Neal and Kobe Bryant and the Los Angeles Lakers in just five games.

I never got into sports, though, and I left White Hall when I was in third grade. With my dad in the Army, we'd move, but I'd come back, and in many ways, I never really left. When I moved with my parents out of the South, guys would try to bully me over my Southern accent, and, though I'd try to keep my cool, if pressed, well, let's just say I wasn't going to let anyone push me around.

Under Pressure

Throughout my ordeal with Pearson, Balch & Bingham, the Tuscaloosa DA, and even some family thought I should take a plea deal and just get it over with. But I could never do that. I wasn't going to let some condescending lawyers from Birmingham railroad me into prison. No, that kind of stuff has happened too much, and, if I could help it, I wasn't going to let it happen to me. Besides, that's not what I inherited and that's not how I was raised.

It's hard for me to pinpoint exactly what made me so defiant in the face of overwhelming pressure. After all, I wasn't even twenty years-old through much of this ordeal, and I was facing the combined efforts of a billion-dollar corporation, its highly-paid attorneys from a law firm with tremendous experience and resources, including powerful political connections, in addition to the Tuscaloosa County District Attorney's office, and the latter eventually secured an indictment against me on multiple counts by a grand jury. It seemed to many people who knew of the case that, regardless of what they knew about my innocence, I was going down. The only questions were how much time I could do and if I might beg and plead with Pearson and Balch to help me avoid incarceration.

I imagine many factors were at play, including my previous experiences. A company had already threatened me with law enforcement when I was in high school at ASMS, and I'd already had companies try to steal my work and ideas. Plus, I became calloused by the tactics that had already been used against me. The more aggressive Pearson became the angrier I became. At one point, I even wanted to go to the media and had fantasies of holding a press conference to expose Pearson's persecution of me. When the UAPD raided my dorm, I couldn't help but think that racism played a part in this entire fiasco, which is why I confronted Dr. Han with such a possibility in our meeting. Combine these with Tankersley's threatening conversation with my mother and him including my grandmother on an email, then you might understand the groundswell of anger that I felt, which by no small measure accounted for part of my motivation to withstand everything that was thrown at me.

But it wasn't just anger. I had to put my anger aside and not let it drive my decisions. Anyone can get angry at what is happening to them, but my anger was buttressed by my sincere belief that I had done nothing illegal. So, I believed that in the righteousness of my cause and that if the evidence was clearly explained and examined, then I would be exonerated, even as my case inched closer and closer to trial. Some around me felt that a trial was a formality toward my conviction; therefore, if it got that far, then I'd be screwed, which is why some on my side were concluding that a plea deal was my best option. If I went to court, the jury could see me as an unrepentant hacker, based on my past that the DA, egged on by Pearson, could resort to throwing before the court.

There may be times when accused people might need to cut their losses and decide that the outcomes of trials and the imposition of sentences are too risky to gamble on. Why take the chance of sitting years in prison, even if you're innocent, if you could take a deal, do no time, and have something better on your record than whatever it would be had you gone to court, forced a prosecutor—who's insulted you didn't take the deal—to do all that work to present the case to the jury, forced the jury to work only to have them convict you? In other words, it might be easier for everyone involved to just admit to something even if you don't believe you broke the law. In a given circumstance, it might be better to take a plea. And that's the choice I had to make. But something told me that I had to take a stand. If I had to take a plea, then maybe I would, but as of the summer of 2018, going into my meeting with Daniel Fortune, I decided to hold my ground.

My sense of right and wrong was what ultimately had the most influence on continuing to fight the charges against me. Part of that sense can be attributed to my own personality, but I also owe much of that to my grandfather, John Jackson, who, along with my girlfriend, encouraged me to stand up for myself if I was sure that I had not committed a crime. No one knows better than my grandfather does about what it means to stand up for yourself in the face of injustice, corruption, and underhanded tactics meant to break you down, humiliate you, and destroy your will to fight. And after reflecting on my resilience against these powerful adversaries, I realized that my strength came from a history of my family standing up for what they believe.

John Jackson

For over thirty years, my grandfather, John, was the mayor of White Hall. He has been the driving force behind what exists in the town. He brought in much needed revenue and jobs in Lowndes County, one of the poorest counties in the United States, by spearheading the opening of the two casinos in White Hall. Neither casinos serve alcohol, and my grandfather will tell you that they're the only places in Lowndes where blacks and whites freely congregate. Over the years, my grandfather had to fight off attempts by powerful state politicians to shut down the casinos, many who didn't think they were getting enough of the cut. He had to fight to keep the casinos open; at one point, one was shut down by the state, and the case made it all the way to the Alabama Supreme Court. While mayor, John was harassed and challenged by state politicians and bureaucrats, having to fight off ethics charges trumped up against him over the years all while trying to bring much needed improvements to the community. This is what happens in many poor, and often black, communities all over America. State politicians won't lift a finger for them for decades, but when a leader like my grandfather comes along and starts bringing in money and other essentials, they all want to swarm in, and, of course, start running the show or start dictating how events are going to unfold. They also want a piece. The state hasn't had a problem leaving behind many Black Belt counties like Lowndes, offering them little, if any, help. In Butler County, for example, south of Lowndes, Newsweek's Carlos Ballesteros reported that "raw sewage "flows from homes through exposed PVC pipes and into open trenches and pits," according to United Nations Special Rapporteur Philip Alston.

Michael Harriot of *The Root* further explored the consequences of these conditions, focusing on Lowndes County in particular. He reported, with evidence from the Equal Justice Initiative, that black residents were arrested when they tried to solve sewage problems that were causing hookworm. The state and local officials, instead of trying to address the problem, criminalized these poor black residents who were trying to save their own lives and the lives of their children. Hookworm is a parasite that was thought to have been eradicated from the United States and the developed world, states Harriot. But, then, according to the UN, areas of Alabama, because of state and federal neglect, does not fit the criteria for belonging to the "developed world."

As mayor of White Hall, my grandfather tried to turn his part of the county around. He brought in essential modern needs like improved plumbing, a new school, and a community center, in addition to economic opportunities, often butting heads with state officials who refuse to invest in counties like Lowndes that really need it. These battles caused him legal problems and eventually led to his ouster as mayor. During his battle with the state over the casinos, the government came after him for supposed ethics violations and other suspicious charges and technicalities. One sympathetic ethics investigator, though, told my grandfather, "Hell, you barely take a salary; how can we charge you with ethics violations?" All this against a man who is responsible for bringing the Lowndes County Interpretive Center to White Hall. The Interpretive Center is a museum of the Civil Right Movement, commemorating the marchers and victims of racial violence as they fought, risked, and, in some cases, gave their lives for voting rights in Lowndes. It's odd that it's not the many state politicians facing ethics charges when they leave many of the counties and people they govern in such sheer poverty and in deadly health conditions that the United Nations must get involved. Where are their charges? Who's holding them accountable?

And this is when the stories that linger in Lowndes start to matter as they pertain to my case. You could tour the county yourself, get a feel for it, visit the sites, read the histories, tour the museum, and then you could begin to understand a source of my resolve. Or you could just ask my grandfather, John Jackson.

He's getting up there in years, but he still has the wits of a young man and an understated sense of humor that catches you off guard. He'll tell you, proudly, that his brother, my great uncle, was the first black deputy sheriff in the county, right before adding, "Yeah, I made him that." Or he might tell you about blacks who get appointed to important positions by whites that they "stay black for about twenty-four hours." But if you're willing to help, he doesn't care what color you are. He's lived it. Give him a call, and he'll be glad to give you a tour of town and tell you about the county. He'll tell you the story and show you the grave marker of Viola Liuzzo, a thirty-nine-year old white woman from Michigan who, in March 1965, was shot nineteen times by the KKK while transporting Civil Rights marchers from Montgomery to Selma. He can tell you how Viola Liuzzo's passenger, Leroy Moton, a seventeen year-old black kid, had to pretend he was dead, so the Klan wouldn't kill him when they came to make sure their victims were dead. He'll tell you about Jimmie Lee Jackson, a black deacon and military veteran, who was beaten to death in Marion, AL by white troopers a month before. He can tell you about Jonathan Daniels a white minister shot by white "deputy" Tom Coleman in Hayneville while trying to protect black teenager, Ruby Sales. There's more, like Rev. James Reeb, and scores of others beaten, shot at, and threatened. These are the better known, well-documented stories, but there's something about hearing it from someone like my granddad who was there, knew many of these people, and was directly involved in the intertwined events. If you ask him if he was there when so-and-so was killed, he'll tell you grimly, "I can tell you about a lot of people who were killed."

If you ask him if he still sees the children and grandchildren of the white terrorists around, he'll tell you he might see them, has even taught them, but holds no grudges against them, except that "many of them still don't vote for black people."

Walk into the Lowndes County Interpretive Center with him, you'll see the employees there know him and respect him and tell you that you are there with the man who got it done. Next to his and my grandmother's house is the Freedom House. And that, truth be told, is where it all got started in Lowndes County.

And here's where you grab a snippet of the stories that hover and flutter in the air like moths to a light in the thickness of the Alabama dark. It's 1965. My grandfather isn't a grandfather or even a father yet; he's, John, a sixteen year-old kid. He's driving a school bus for work while going to school. He sees some Civil Rights workers coming through and stops to listen to what they have to say. He'd grown up in Lowndes, and his daddy had to teach him that when they head into Selma or some other town in the state that there are white and black water fountains. And he grows up seeing the violence of racist whites against blacks. He's grown up in fear, but also with a chip. He's got fear because his and any black person's life is in danger. But it's not a fear that would discourage him or prevent him from standing up for himself. He's also been out in the fields. He picked cotton like generations of the family had done. Imagine being in that heat, sun on your back or in your face, and your hand twisting the ball out of the boll and doing that for hours and hours with your back bent over and dry dust whipping up in your face if the wind came through. If your mouth is open, and it might be because you're thirsty, that dry dust peppers your tongue and you breathe it in, and that powder soaks up whatever moisture is left in your mouth. You do it because it is what you must do to help yourself and your family.

Family History

John is the youngest of ten children. He's the son of Mathew and Emma Jackson, who became among the first independent black landowners in Lowndes County when the federal government divided up the remains of the White Hall plantation. After growing up working the land, John was offered the job to drive the school bus at sixteen; the county offered him the position at such a young age so it could save money. There weren't many jobs or economic opportunities for blacks off the plantations, so he was eager to earn money that didn't involve working in the fields. In fact, so many blacks left Lowndes to look for work elsewhere so they didn't have to pick cotton that the county's population was in free fall for decades. The US Census records show that the county had a population of about 32,000 in 1910. By 1960, that number was reduced by half (the population as of 2019 stands at about 11,000).

The stories about my grandfather and great-grandparents have been recounted many times in documentaries and books about the Civil Rights Era. My grandfather has been interviewed about the events that surrounded Bloody Sunday in Selma and that tumultuous year of 1965 when the high profile cases of KKK killings made headline news around the world. But the racist violence in Lowndes has a long history, and just like my grandfather has said, many people have been killed over the years because of racism.

The violence and intimidation and the existence of the Klan created, of course, a heavy atmosphere of intimidation and terror that was just as much part of Lowndes as the cotton that grew in it. Blacks grew up with it as if it were attached to their DNA. The Klan could come at any time, but they came as cowards in the night, and they terrorized white sympathizers and helpers of blacks with just as much ferocity as they did blacks they deemed troublemakers or even just those that started to gain some power in the community. Because Mathew and Emma had become known as landowners in the area, they could have easily become targets. In the 1940s, a black man named Elmore Bolling was lynched by whites, shot six times, after becoming a successful businessman, and that weighed on the whole Jackson family throughout the era.

But the Klan's reign of terror didn't deter John Jackson or my great-grandparents. My grandfather remembers that when he was sixteen the Student Nonviolent Coordinating Committee (SNCC) came to Lowndes trying to get students involved. He's revisited this account in books like Bloody Lowndes: Civil Rights and Black Power in Alabama's Black Belt by Hasan Kwame Jeffries, Circle of Trust: Remembering SNCC by Cheryl Greenberg, and in documentaries like Eyes on the Prize II: America at the Racial Crossroads 1965 to 1985. He's also told these stories to me and anyone who has asked to hear them, though he's done more in his life since the 1960s as he continued to battle the entrenched regressive political policies that unfortunately dominate Alabama politics. I mean, how hard is it in the 21st century to make sure that the state's counties don't have an outbreak of hookworm? Or that black neighborhoods aren't poisoned by pollution by companies protected by law firms like Balch & Bingham?

Nevertheless, in 1965 sixteen-year-old John Jackson was making fifty-dollars a month, which is about $400 a month today. Then these SNCC workers started appearing in Lowndes, visiting schools, meeting with students, and trying to study the quality of education blacks received in the county. One of the schools my grandfather attended still stands. They had to squeeze kids from different grade levels into one small building. John eventually attended Lowndes County Technical School, for which he drove the bus, a black school that, though an upgrade from the average black school in the county, lacked the resources of white schools. John started to pay attention to SNCC and let them on his bus to distribute SNCC literature. SNCC hoped that the students would bring the flyers to their parents and start getting them involved in the voter registration drive.

Decades later, if you were to visit White Hall, you might stand in the heat before the Freedom House. My grandfather might be there with you, if you ask politely. You might stand next to him in the freshly cut yard, and if you go in the summer you will feel the weight of the heat pressing down on you, a heat so thick and moist that even the thinnest clothes start to sag and stick to your body. If you look someone in the face in that heat, they will appear to be melting before you, their brows streaked in sweat, their arms bubbling, and shirts darkened in wet patches. Bees and wasps might dip in among your group, if there is one, while yellow butterflies flap past you. There may be no sound but the chirp and call of birds; the grounds are peaceful, hallowed, humble. The scent of pine and grass are baked into the air. The Freedom House is a white house, but its architecture can make no claim to grandeur. It sits, unevenly, alone, a tilting rectangle. Raised on bricks, its metal roof is ruffled and spotted with rust. My grandfather might take you inside, but before you go, you might imagine this man in 1965 as an exuberant sixteen year-old kid, driving a school bus, and coming home to tell his family about the strange people he met on the road. And his parents with calm and consideration would listen.

They would listen and know the significance of what their teenage son has told them.

Many people after that day—the day that John Jackson came home to tell his parents that the marchers had come to Lowndes—would not, in my grandfather's words "fool with" the Jackson family, meaning that they would no longer want to associate with them, because they knew it would mean trouble. Mathew and Emma knew it, too, and through them, John would know it. But it did not deter them. If they were afraid, and they must have been, they did not let it stop them. The rippling effects of the Civil Rights Movement had finally reached this thinly populated, agrarian county, and Mathew and Emma realized that the time had come to catalyze the changes that were afoot.

That time may seem distant, severed, inconsequential to the present in which we live, no matter how much, with our sense of history, we know we should care, an obligation that seems more forced than naturally moral, an unspoken scold that presses upon the conscience. More than half a century arrives like the light of stars carrying in it a record of the past. The younger minds, generations removed, may look around, think of their own lives, their immediate opportunities and lack thereof, or they may not think of any time but the present, the moments, or beyond their own lifetimes, and they may see nothing, if they are to stand in front of it, but an old house; someone tells them, and maybe they interpret it as such even if it is not, they may interpret it as a lecture, something that they should care about but cannot make themselves care because they do not comprehend its immediate relevance to them nor do they look around and find in it any immediate utility. They may never feel it, sense it, or understand it in any real way, and its significance escapes them even though it has protected and nourished their lives, bodies, and souls. It has provided them with the dignity from which they may act brazenly, openly, and choose to live another type of life should they imagine it.

More than half a century removed, the house, the marchers, my sixteen-year old grandfather there may seem no relevance to me, Pearson, and Balch & Bingham. And perhaps I did not sense it, feel it, know it myself at the time, or always think of it, but there is something not-just-coincidental about the bravery of my teenaged grandfather, my great-grandparents, and others who stood up to forces trying to compel them to surrender their pride, dignity, and self-worth to those who care nothing for them. I have been asked why I did not give in, why I did not accept a plea, why I did not cower even after having been arrested, humiliated, and threatened. In my conversations with Dr. Love about my ordeal, he told me that he had discussed my case with a man from Lowndes County, who had grown up there, went to school there in the aftermath of the Civil Rights Era, and who now owns land there. He, a white man, asked Dr. Love about my reaction—this African-American kid from Lowndes—my reaction to the charges against me and the attempts to bend me to their will. Dr. Love replied to him with something like, "He pretty much told them to go shovel dirt." And the man laughed and nodded his head and answered back, "Sounds about right."

It would be wrong to assume that blacks in Lowndes County were meek before the Civil Rights Movement. After all, at the beginning of the twentieth century, a black man, Jim Cross, was lynched along with his wife children after speaking out against the lynching of another black man. In 1917, in the same town, Letohatchee, which has only ever had about a hundred people, two black men, William Powell and his brother (either Samuel or Jesse) were lynched for being disrespectful to whites (according to the state historical marker). Others were killed, shot at, terrorized for being outspoken or too successful, like Elmore Bolling. His story is recounted in his daughter's book The Penalty for Success (by Josephine Bolling McCall). The problem was never that the county lacked blacks who stood up for themselves and others but a problem of real physical and psychological terror that helped to prevent organization and more vocal resistance. The arrival of SNCC, therefore, galvanized an already existent pride and strength. In Bloody Lowndes, Hassan Jeffries records that SNCC's, especially Stokely Carmichael's, refusal to kowtow to white deputies was a watershed moment for the blacks of Lowndes County. Thus, part of SNCC's legacy, in tandem with the blacks of Lowndes and some white helpers, was the eradication of racial terrorism and the fear it spread. Over time, a new defiant, rugged, gritty attitude was bred into the subsequent generations. It's why Dr. Love's confidante quipped "Sounds about right" when hearing of my reaction to the charges and demands. Many of the people of Lowndes are now known for their recalcitrance, and, as I mentioned before, it can be a rough place. That's why my grandfather would caution that you still need a moral, optimistic, forward-looking base for your attitude to be progressive and productive. Otherwise, you just have misdirected anger that does more harm than good. If you have nothing morally good to gain for yourself or anybody else, then nothing is exactly what you'll get.

So, there was the sixteen-year-old exuberant John coming home to tell his parents about SNCC, handing over the leaflets, wide-eyed, heart thumping. And maybe they sit for dinner and the sun has set and the day's light has dimmed. And the prudent parents take the literature, and they now realize the time has come, and maybe their hearts thump faster but the blood that surges through their chests is mixed with ambiguity, fear, and uncertainty that comes from maturity and experience. They have lived long lives; they have seen and heard much more than their youngest child. But they take the leaflets, and they study them; maybe they hold them in their hands that have toiled in fields, dug in the earth, built homes and furniture, held and protected children and grandchildren, and hands that have written their names to register to vote. Coincidentally, they have already paid a price. Their registration has become known, and a local store owner, in retaliation for their daring to think that they have such a right as voting in an election, has demanded full payment of all debt. Mathew and Emma pay the debt but vow never to do business with the store again. John is fired for letting SNCC material on his bus. These incidents are true; my grandfather has told the story many times to many people.

But they do not back down. My great-grandparents had great hopes for their children. They believed in education, and they made sure their children would go to college, as John would later do. They had made sure that their children were intelligent and aware of what was happening around them, and they recognized that John's enthusiasm for SNCC was the natural result. They knew that education is about what SNCC was trying to do: create an informed voting citizenry who could self-govern, the very essence of a democratic republic. He's been quoted in several records, and I will restate the quotation here: in response to the looming danger his family faced for deciding to take a stand, my great-grandfather, Mathew Jackson said, "If we're not for ourselves, who can be for us?" By all accounts, Mathew and Emma became like parents to the activists of SNCC. At John's suggestion, my great-grandparents opened a family house for SNCC, which is now known as the Freedom House. SNCC members lived there and used it as a base of operations as well as a rest stop between Montgomery and Selma.

Terror At Night

We can imagine the lamplight in the house at night glowing against the wooden walls. Just outside, the dark night grows darker, and the view from the sturdy porch reaches only a few yards before vanishing into a bleak void. Lights in the yard come from cars coming or going, but once they are gone or the driver has parked for the night and shut off the headlights, the great shadow renders the viewer almost blind. Voices greet each other, and the click of shoes go up and down the steps, and the porch bellows from the weight of men and women moving across its planks. Films of sweat rests on top of exposed skin, and the night air smells of wet trees. Once all are inside, a man tries to keep watch from the window. He tries to see the road, but it's too far, so he waits nervously. He is caught between wanting to see lights, hoping that if compatriots are tired or lost or have come to join them have found their way to the home, and not wanting to see them for fear that what emits them is a truck full of armed Klansmen.

The nights pass quietly, though there are restless sleeps. With each night, conversations, discussions, and laughter fill the house and seep into the walls, and if you visit the house you might hear the remnants of such escaping from the cracks and crevices. While exchanging stories of their adventures and histories, they plan strategies for registration, defense, protest, and teach John about their views. The nights, though, grow more eerie. The longer nothing happens, the denser the tension and anxiety become.

During the day, people are being evicted from the plantations for registering to vote. Tents go up in a field off the highway to shelter them. There are plans to move more onto Jackson family land. Entire families are housed in sheets that over time are torn by wind and rain. At night, white terrorists drive by and shoot wildly into the tents. The blacks, though, are armed, and they are not afraid to fire back. Even children stand guard.

We might imagine such nights but can never fully relive their terror, not in the sense that reliving is to relive it with fear and uncertainty or to relive it with the pain and anguish of the murdered and wounded. We might imagine the silence of the night settling in until it blankets us, and our eyes grow heavy, wanting to shut, so that our bodies can sleep. Others who keep vigil cannot bear to feel tired; each sensation of fatigue renders them with guilt. There are lamps and fires shining yellow spheres tinged with orange into the darkness. Footsteps pat the ground from restless watchers. Minds and souls grow heavy with apprehension. The worries are not limited to the night's potential terror but stretch into an oceanic future. You can only be sure that there is something beyond this moment, maybe not for you should you not survive the evening, but there is a future for most around you. And like staring outward into the sea in which the only certainty is that it is there and that it is long and endless and that there is something to go into, you know only the incalculable dangers that lurk in its infinite waters. Maybe one of us goes to bed, slinks into a cot, trying to decide whether to cover our body with a sheet or to lie bare. We hear the mumbles of voices from outside, and once in a while we hear a cry or a laugh or the heart-and-ear-breaking shrieks of babies.

Amid the campfire and lamplight men and women speak. Their voices crisscross and exchange bits of each other. The men's faces are coarse with hair. Some shut their eyes, and you try to determine if they are asleep or if they dared to let their minds go blank for just a moment's rest. An old woman rocks in front of the slit of her tent. A rifle lies across her lap. A boy next to her on the ground, one who refuses to go inside, wonders if the old woman has ever shot a gun in her life and what it might be like to shoot one. Light singing reaches your circle from somewhere else. Bugs tickle your neck as if they went for a dip in the sweat that covers it. The conversations move you, and though you cannot forget the horror of the circumstances, something about them has brought you closer to these people whom you now regard as brethren.

The wood crackles from dying campfires, and someone from a few tents down hums a hymn. You wonder how many are still saying their prayers, and you wonder if you should pray again. The morning must come. Daylight banishes the terrorists like God's glory extinguishes frantic demons. You linger and linger on the edge of sleep. You drift, and you're into sleep, and thankfully you have no nightmares. And the sun finally rises, and the camp stirs with the optimism and relief of a new day. You know, though, that because nothing happened that night, then something is more likely to happen next. So, it does. One night, a night that has become family and county lore, the terrorists come for the Jackson family.

The sun withdraws itself. It tucks itself beneath the horizon. Lamplight spreads weakly across the porch. Moths and flying specks flutter and zip from the light to the dark and back again. Country night, country sky. Beyond the country homes and tents barely a living soul can be found. Miles and miles of empty land lie in still waves. Out there are only the primal sounds of hidden creatures. In the stillness of the evening, in its loneliness, in its country solemnity, anyone might wonder what is to be fought for and defended in this deep trench of the earth. The simplicity of life in this county demands no such importance to have the ferocity of enemies descended upon them. For a century, the people who have lived here have chosen to live in the most basic ways, with no demand for little more than farm, house, church, and family. They have lived with little more than a rudimentary expectation of justice, honest dealings, and a simple notion of choosing officials with whom, ironically, they would rather not have much to do with. The greatest percentage of their lives are spent with relatives and neighbors in labor and worship.

Yet, in this random place, a young man sits on the porch with a rifle in hand, wanting to sleep. He has heard rumors that terrorists will burn down his family's house because his family has provided people who have no homes with the pride of having somewhere to go, somewhere to live, even if only temporarily. And he sits on the porch and waits because his father has told him to be on guard. Of all the places on earth, of all the materials in the world, of all the seats of power, a group of people have seen fit to wage war here, in a tiny hamlet, that from not too far above lies invisible to the world. We can imagine the teenager's hands shaking as he tries to steel himself in this baptism-by-fire of manhood. And he knows it, the way some adolescent boys know that they are facing the watershed moment of their lives, yet it is not a practice run; real lives are at stake, including his own. It is not a question of being scared or not being scared for fear is natural, inevitable, and even useful. Fear is what makes him keep looking out, to notice every sound, every hint of light in the distance. He peers into the darkness and waits. After a while, after sitting still and stiff, he forces himself to move. The first twitches are frightening as if his movement might summon his attackers. He walks across the porch; his footsteps and the creaks are the loudest sounds in the county. Glancing around the corner of the house, he grips the rifle tighter. He peeks into the nothingness and returns to the center of the porch. His body relaxes and slinks to the steps where he takes a seat. His fingers loosen, and the rifle lowers to his lap. This could be a forever night, and so could the next one and the next one. He wonders if he will ever sleep again.

The tree line looks like the rim of a hole in which he is at the bottom. Into this pit comes starlight and moonlight, but they do not reach the ground. The fear is no longer constant. It recedes and returns. He leans and shifts to find a more comfortable position. The edge of the step sticks into his back. He scoots up and lifts his head. A light hits his eyes, and he freezes like a deer. They've arrived. His heart plummets to his stomach. His eyes squint. A series of shots shatter the night air. He expects to feel pellets or bullets rip through him. He doesn't. Maybe they've passed through him and in his shock, he hasn't felt it. He gathers himself, adjusts the rifle, points it toward the lights, and fires. His father has heard the shooting and returns the fire. The vehicle speeds off. Mathew tells John to go up the road to John's sister's house to make sure she's alright. He gets into the car, and the wheels rip the dirt from the ground. He catches up and tries to get the license number on the car. Four men are shooting into his sister Dorothy's house; she was a schoolteacher, but she also had helped bring weapons to the Civil Rights workers because she knew there would be violence. They turn fire on John's car. Bullets, seventeen in all, smash into the car, and then the terrorists speed off, vanishing into the blackness.

From the partial tag my grandfather was able to give them, the FBI caught up to the men. But the agents told him the men just said that they were hunting. Sure, they were hunting black folks, my grandfather liked to say. Despite the threats and violence, my grandfather and his family were not about to back down.

Original Black Panther Party

In 1966, my grandfather was part of the creation of the original Black Panther Party, or officially the Lowndes County Freedom Organization. The party was created because neither the Democratic nor the Republican Party would allow blacks on their tickets, so they created a new party and new symbol, especially to let illiterate voters easily distinguish among the candidates' political parties. Other civil rights leaders used the symbol and created the more famous organization. My grandfather quickly integrated into SNCC, went to Michigan, and then traveled to Russia where he met Soviet premier Nikita Khrushchev. He also met with Muhammad Ali and later with President Ronald Reagan. As much as he traveled and accomplished, my grandfather returned to Lowndes County to become mayor and try to help the county rebuild and bring it as close to the twenty-first century as he could manage. When he became mayor, he secured for the town incorporation, which made White Hall eligible for much needed federal funds along with other basic services. He even used his own money to cover some of the town's expenses. Stories about my grandfather's generosity and desire to help the community are well-documented in various news stories, interviews, and accounts from the town's citizens.

Decades have passed, and the struggle of the county continues. Water sanitation continues to be an issue as county and state officials still neglect towns like White Hall. There are little towns like it all over Alabama that are begging for help for twenty-first-century medical services, clean water, education, and to make sure that lands aren't polluted, or wanting help to stop selling to companies that will pollute them. Politicians neglect these places, though, unless there is money to be made, and then they are only willing to help themselves and other powerful private-but-well-connected entities that can swoop in and take advantage for themselves.

Benevolent political action is a rarity. The people of these towns are left the task of solving million-dollar problems with nickels and dimes. For the newer generations, they pay for government officials' derelictions. The options for young people in Lowndes are as limited as the state's politics has made them. Ten years after SNCC left the county, the KKK resurged and tried to reestablish its dominance and had to be beaten back again. The remnants of those racist times persist but in more subtle ways. What was once direct acts of violence has now morphed into indifference, virtual abandonment, and more sophisticated corruption. The populations are either declining sharply or flatlining, and if they continue to do so, such decimation or stagnation legally justifies local, state, and federal refusal to invest even humane services into these areas.

On top of that, because rural places with small populations lack money and political power—their small populations leaving them with little representation—they are vulnerable to predation by corporations or companies like mining, timber, and processing factories that are represented by law firms like Balch. If such a company is from out of state, it can use an in-state law firm and its connections to sway the legal and political balance toward them, even if the company has a damaging environmental impact on communities.

I can't say that I was thinking of all this county and family history during my battle with Pearson Education. It's not something that I would have found necessarily relevant to the specifics of my case. I was not involved in a life-and-death struggle with armed racists firing into my house, nor did the fates of thousands of people lie in my hands or lie in the balance as my case came to a head. But what was at stake was my freedom and my future. I had to ask myself what I was willing to risk by standing up for myself. I could see, upon reflection and from my grandfather's belief in me, that much of my spirit descended from my family's past. They were and are strong people who had endured and persevered, and I was taking part in that tradition. Because my family had risked their lives doing what they believed was right, I had the strength to do the same. It was time to steel myself, not for prison, but for a long battle.

CHAPTER 7

Pearson and Balch & Bingham

To counter Pearson's actions, David Schoen and I decided to do some investigating of our own into Pearson Education. Based on their actions and demeanor of their attorneys, we believed that there might be a pattern of behavior with this corporation. Plus, by now, I was certain that Pearson was mostly afraid that I knew that its system left exposed the confidential information of millions of students. By seeing to it that I was convicted of a crime or crimes, Pearson could maintain that its programs were adequately secure but had only been breached by a maliciously illegal attack perpetrated by an experienced and well-documented hacker. Therefore, Pearson would be able to mitigate its responsibility at the least, if not totally exonerate itself, should it face an inquiry about the security vulnerabilities of its software. Pearson wasn't going to risk lawsuits or losing contracts with universities all over the world without doing whatever was necessary to see to it that it was put on record that I was a criminal who had targeted and smashed into its programs. If it meant watching me get marched off to prison and destroying my life, then so be it.

Stalking Minors

David Schoen and I investigated news reports and court cases involving Pearson Education, and we discovered a history of unethical and illegal behavior by the British-based publisher. In March 2015, Bob Braun, a former journalist for Newark, NJ-based The Star-Ledger, reported on his blog that Pearson Education was monitoring the social media accounts of high school students, or in other words, of children. Braun's reporting was substantiated by several news outlets, such as The Guardian, Fox News, and The Huffington Post and other subsequent investigations, and even Pearson admitted that it was, in fact, monitoring students' social media postings to detect cheating. This story sounded all too familiar to me because it involved Pearson delving into the online postings of minors, just like Pearson had culled some of my postings before I had become an adult. In a March 22, 2015 interview with Perry Chiaramonte of Fox News, Mr. Braun stated that Pearson was "seeking out the bad guys and punishing them" to avoid the costs of having to remake tests because students might post test questions online. Braun's scoop went viral and educators and parents expressed outrage and concern over Pearson's invasion of children's lives and online conversations.

My attorney David Schoen reached out to *Huffington Post* and *Daily Kos* writer Dr. Alan Singer, who also is a historian and professor of Teaching, Learning, and Technology at Hofstra University. Dr. Singer had also reported on Pearson's monitoring of high school students' social media accounts as well as several other stories involving Pearson, including having written a June 2017 article titled "Pearson and the Neo-Liberal Assault on Public Education" that documents Pearson's profit-driven business strategies that come at the expense of quality education. The article summarizes Singer's larger essay of the same name, co-authored by Eustace Thompson, that provides further detail about Pearson's profiteering maneuverings. According to Singer and Thompson, Pearson and its for-profit-school-partner Bridge International Academies were "forced to end operations in Uganda" and BIA was under investigation by the British Parliament. Such efforts by Pearson have been part of a larger strategy to profiteer by invading Third World countries and partnering with other corporations under the Trojan horse cover named Project Everyone to influence such governments' educational policies in its favor. This new corporation co-opted a United Nations' initiative to provide "free, equitable, and quality primary and secondary education" into a money making venture; Project Everyone, according to the essay, "copyrighted the icons and summary titles being used to promote the United Nations initiative." In his HuffPost article, Singer also documents Pearson's financial setbacks as a result, in part, of growing "hostility to Pearson's brand." As late as February 2019, Singer wrote an article for Daily Kos titled "New Mexico Declares War on Pearson Mis-Education" in which he asserts that Pearson was "cheating [New Mexico's] students and taxpayers" by setting up a for profit school through a subsidiary called Connections Education. The school closed because "it failed to properly educate students." Starting in 2020, according to Singer, New Mexico "will join Maryland and New Jersey in no longer using" Pearson's PARCC exam.

Dr. Singer expressed interest in my story at the time Schoen reached out to him, but at that moment, legally nothing was settled, and I did not know all Pearson's machinations behind the scenes. Singer, though, revealed that Pearson, concurrent with my case, was losing or had lost contracts with states, including Texas, worth hundreds of millions of dollars. Although Pearson's troubles might seem tangential to its battle with me in Tuscaloosa, knowing Pearson's growing woes, I began to understand its desperation to keep its security lapses under wraps for fear it might cause other schools and states to stop using its products.

Losing Millions

In July 2015, *The Washington Post* reported that Pearson had lost a multimillion-dollar testing contract in New York. The article, written by Valerie Strauss, went on to report much more of Pearson's financial bleeding because of the inefficacy and incompetence of its educational testing and materials leading to an exodus from its services by many more states. In Indiana, the paper reported, Pearson's lack of security for its computerized exams came under scrutiny.

As my lawyer David Schoen and I dug in further into Pearson Education's history, we found that Pearson had a long history of behaving unethically, immorally, and illegally. Though he practices in Montgomery, Mr. Schoen also has a law office in New York, so he is even more familiar with Pearson's legal troubles there. Of course, any large corporation is going to have its share of legal issues and lawsuits, but not all legal problems are equal. Some result from deliberate, calculated actions that are criminal. In 2013, Erik Sherman of CBS News reported that New York settled a $7.7 million-dollar-lawsuit with Pearson because of the latter's alleged use of its charities for funneling money to its profit-seeking endeavors. In one example, Sherman writes that Pearson would set up international conferences under the guise of its charity as fronts for sales meetings with education officials. Additionally, Pearson used its charity to strike deals with the Bill and Melinda Gates Foundation that would be steered towards its profit-making operations. According to then-New York State Attorney General Eric Schneiderman, this was a violation of New York state law. In February 2015, Stephanie Simon of Politico in her article "No Profit Left Behind" wrote that "Pearson makes money even when its results don't measure up." Simon revealed that Pearson used "a free online social learning network" called Project MASH that was a massive personal data collecting tool that gathered "names, phone numbers, email addresses and dates of birth as well as information gleaned from [participants'] responses to surveys and their posts in online forums." Though Simon notes that Pearson allowed participants to opt out, she adds that "the privacy policy also let the foundation share users' personal data with Pearson 'for business and operational purposes related to [Pearson's] mission.' There was no opt-out provision in that clause." Simon's article reminded me of Pearson's controversy in New Jersey in which it had monitored the social media accounts of children to gather information about what they were sharing. The purview of its "business and operational purposes," of course, remains vague and to what extent that Pearson uses such personal information is anyone's guess.

In January 2015, *Fortune* detailed (in an article titled "Everyone Hates Pearson") Pearson's disastrous deal with the Los Angeles Unified School District. Accused of using conferences to persuade education officials to support its bid, Pearson, along with Apple, provided iPads "loaded with Pearson education material" to 650,000 schoolchildren. The conferences prompted an FBI investigation, according to the article's author Jennifer Reingold, though the US attorney declined in 2017 to file charges.

It wasn't lost on me that many of Pearson's problems were taking place just before or during the time the corporation was coming after me. Its educational products were being criticized all over the country for their failure to deliver on Pearson's promises of providing effective education. The negative publicity exposed Pearson's greedy attempts to dominate the US and international market, its underhanded tactics to win government contracts, and shady dealings with its Africa initiatives. Another aspect of these stories that caught my eye was Pearson CEO James Fallon's statement in April 2018 to Edsurge, an education technology news and resource site, that Pearson was investing $1 billion in its research and development of mostly digital products. I thought of my case—that it could—and should—compromise customers' faith in Pearson's protection of their private information, which was left exposed in its MyLab software. With such a massive investment, Pearson could face a tremendous loss if the details of my case were publicized.

One of the most frustrating aspects of my ordeal with Pearson came from my grappling with Pearson's persecution of me when I was thinking it was a corporation that cared about education and the students that it served. I wondered why a billion-dollar conglomerate would come after an eighteen year-old with such vicious and dogged pursuit, even after I had taken my website down in June 2016. Indeed, I sent Mr. Tankersley an email telling him such, and naively, I thought that this would end the whole matter. Tankersley's response was to make further demands and threats and that's what I could not wrap my head around. By then, Pearson knew my age and knew that I was a student, so I thought because it was supposedly a student/education-based business Pearson might treat me and the incident with a more circumspect response. Perhaps its representatives could have offered a reprimand and some guidance to an eighteen-year-old college freshman. At the time, I had the same thoughts about the University of Alabama, and even went to the provost to ask why the University would treat one of its students this way regarding the involvement of the UAPD. While the University came around, Pearson dug in. I couldn't understand why it wanted to see me prosecuted. Of course, I knew at least a major component to the answer to that was Pearson's desperate need to protect itself from lawsuits, its contracts, and its ability to assure schools that its products were secure.

Really About Education?

But as I understood more about Pearson Education, I began to realize that just because it has the word "education" in its name doesn't mean that education is its primary concern. I knew, of course, that Pearson wanted to make a profit, but many businesses want to make a profit while caring about the humane value of its products and services and the people who help make it successful. Although students often have no choice of what education companies use, college students are better positioned to voice and critique the products they are using than elementary and high school students. So, I thought Pearson might have a better appreciation of and respect for college students as customers, primary consumers, and those that offer Pearson and its clients valuable feedback on the quality of its materials. Perhaps Pearson itself might learn from me how to better protect its system, and the two of us could have learned from each other from this experience without the threat of criminal charges and lawsuits; perhaps that would not be too much to expect from a business that is supposed to care about education. Plus, I thought an education-based company might want to be a good steward of public trust and be a good citizen in the areas in which it operates.

Dr. Singer had been investigating Pearson for some time. As early as 2012, in the aftermath of Pearson's charity scandal in New York, Singer wrote an article in the *Huffington Post* titled "Pearson 'Education'—Who Are These People?" in which he explored the careers of the three chief operating officers of Pearson at the time: Glen Moreno, Dame Marjorie Scardino, and William Ethridge. Moreno, Singer discovered, had no background in education other than he graduated from Harvard and was mainly an investment banker who lacked credentials as far as understanding of educational pedagogy or ethics. Scardino had been a lawyer and a newspaper publisher, and Ethridge had worked at Prentice Hall and Addison Wesley. None of these individuals had any real pedagogical education experience or expertise in terms of what works in classrooms at the various levels of education. Singer concluded that these individuals were "busy trading stocks and racking up dollars and pounds while the corporation's financial situation is shaky." Closing his article, Singer asked, "Are these the people we want designing tests, lessons, and curriculum for our students and deciding who is qualified to become teachers?" (Pearson designed tests for teacher qualification which came under scrutiny in Florida).

Although Pearson's leadership changed in 2013, I started to think about Pearson's leadership during my case. John Fallon had become CEO, and in doing some research on him, I discovered not much had changed about Pearson's profit-above-education model for which many have criticized the company. After all, Fallon was the CEO during other Pearson scandals. I noticed that in interviews that Fallon has given, he rarely discusses the specifics of education. He speaks of billion-dollar investments and business strategy; true, he mentions getting feedback from students, parents, and educators to make better products, but Fallon doesn't demonstrate any interest or passion about the vocation and pedagogy of teaching and learning, despite paying lip service to such. When confronted with educators' and parents' criticisms of Pearson's education models and products, Fallon deflects such criticism and shifts into banal corporate-speak about trying to do better and learn or pass the buck onto politicians. For example, in reply to criticism of Pearson's standardized tests, Fallon washes his hands of responsibility. According to Forbes, Fallon "suggests" that if educators and parents have a problem with standardized testing, they need to "take it up with the politicians," and, this is a direct quotation from Fallon, "It's a matter for elected representatives." I found this an appalling statement considering that Pearson lobbies politicians and educators to use its products. Second, just because politicians make bad choices regarding education doesn't mean that Pearson must indulge them. If Pearson and Fallon cared about children receiving a quality education, they could choose not to participate in an enterprise that they knew from empirical evidence and expert testimony is harmful to student learning. In short, just because politicians use a bad product on helpless kids doesn't mean Pearson has to supply it.

Such a response by Fallon complements Pearson's history of using its power in the education business world to drive up profits even if its educational products and models don't work, as demonstrated from New Mexico to Uganda. In 2013, London's Evening Standard reported that immediately after taking over Pearson, Fallon shut down Pearson's UK adult-learning program because it wasn't turning a profit.

As late as May 2019, Pearson-run for-profit schools, under Fallon's leadership, continued to fail. In her article "New Study Gives Failing Grades to Virtual Schools," appearing in the *Detroit Metro Times*, Suzanne Potter reports that Pearson's virtual charter schools in Michigan were not meeting performance standards because students were completing "their courses less often, and have lower scores on standardized tests, compared to kids in brick-and mortar schools" based on research of the National Education Policy Center.

The article quotes Michael Barbour, an author of the research, who blames the for-profit model, claiming decisions of the schools are not made "based upon instructional design" but "what is the most cost-efficient way" of running the school. Barbour also states that attempts to regulate these schools are countered by "lobbying by the online schools," which includes Pearson Education. The article, therefore, provides evidence that Fallon continues Pearson's history of placing profits over performance to the detriment of students.

Another point of criticism of Mr. Fallon is that executives were receiving pay raises while, according to The Financial Times, Pearson was suffering losses. The Times reported that in 2017, "shareholders voted to reject a pay increase on one of that year's biggest revolts over executive pay in the UK." These executive pay increases were implemented when Pearson was in the middle of cutting jobs to slash costs. Moreover, The Times points out that shareholders were upset about "a pension contribution equal to 40 per cent of Mr. Fallon's salary" that included "a benefit scheme [that] has since reached its maximum level of contributions, and while he received no further benefits under the plan, he continues to accrue a cash supplement worth 26 per cent of his salary." Pearson did experience a slight turn upward, but in January 2019, CNBC and the Times reported that Pearson revenue had decreased and that sales in the US were declining. Nevertheless, Fallon received a 1.5 million pound-bonus in March 2019. This Is Money, a British-based financial website, wrote that Fallon received retirement payments at a higher percentage than Investment Association guidelines advise. Despite Pearson's reporting of profits, many investors see these profits mainly because of cost cutting rather than real growth; Bloomberg stated in January 2019, that Pearson was "analysts' lowest-rated stock" and that "[b]rokerages see no end to Pearson's struggles."

Pearson and Fallon, however, remained ambitious. With US sales faltering, Pearson has aggressively continued its forays in the global market and international education. In April 2016, the month of UAPD's raid on my dorm, Anya Kamenetz published the article "Pearson's Quest to Cover the Planet in Company-Run Schools" in Wired magazine. The article details Pearson's teaming up with the Philippines-based company Ayala to spread for-profit schools around the globe. The company is called APEC, using the slogan of "Affordable World Class Education from Ayala and Pearson." Pearson, Kamenetz notes, is trying to take advantage of the annual $5.5 trillion-dollar global education market by testing its "academic, financial, and technological models for fully privatized education on the world's poor." These schools charge a low fee for attendance, fees that are often placed on the burden of taxpayers. Therefore, Pearson is guaranteed to get paid either through individuals who can afford the fee or through the form of government vouchers for those who can't. Furthermore, Pearson intends to continue to use standardized testing to measure students' performance and progress. Pearson's move sparked an international response from "major teachers' unions in the US, UK, and South Africa" that wrote and signed a letter to Fallon decrying these efforts to undermine free education. Such criticism of Pearson's profiteering in developing countries led to a United Nations Human Rights Council resolution "that called for monitoring of all private education providers."

Pearson maintains that its low-fee tuition provides more poor children with a chance at a better education. While in some instances that may be true, according to Kamenetz, these low-fee for-schools have a history of failing and exacerbating problems in poor countries rather than solving their education problems. For example, many poor people still cannot afford the tuition and poor people who can pay may have multiple children. They may be able to afford the tuition for one child but not for others. Second, when parents are faced with such a choice, "boys are educated in favor of girls," creating even more sexism than already exists. Plus, although some poor may be able to afford to pay, the most likely results are that wealthier kids end up in such schools, furthering socioeconomic class divisions.

Fallon placed his trust in Sir Michael Barber, who has been called the most influential education expert in the world. Barber became the Chief Education Advisor of Pearson in 2015. Despite his accolades, Kamenetz reports that Barber has his share of critics such as David Archer of ActionAid, who helped stop previous plans to charge poor children for school because it led to increased inequality. Another critic of Pearson's plans is Jishnu Das who has criticized Barber's strategy in countries like Pakistan. Eventually, these critics argue, there becomes a self-serving slippery slope to Pearson's for-profit school strategy. Because Pearson charges tuition, they are bound to steer millions of children away from free education schools and by convincing governments to provide vouchers, the governments divert badly needed funding from its public schools. Once these public schools collapse or become further marginalized, Pearson will corner the market on education in these countries, leading to even more financial windfalls; however, the poorest children will be left behind in its wake. Finally, Pearson's measurement of success and learning will come from its own tests and products. Would Pearson admit that its own products and methodologies fail with so many billions invested and so many billions to be earned, even if the schools do not perform well? Who will know but Pearson? Who will hold Pearson accountable? Kamenetz asks. Moreover, whatever virtues Berber may have, he left Pearson in 2017.

The Washington Post picked up on Kamenetz's story and furthered the investigation into its international operations. Part of Pearson's motivation to move aggressively internationally has been more and more hostile reaction by US educators over Pearson's influence in American education. The Post's April 21, 2016 article by Valerie Strauss cites a scathing letter from the Chicago Teachers Pension Fund and UNISON, a British trade union, as an example of increased and high-profile criticisms, with some of those criticisms focused on Pearson's international initiatives in poor countries, its educational methods and products, and the degradation of its reputation in the United States. Exhibiting hubris, Pearson dismissed these concerns in a response printed in the Post, which prompted a critical response of Pearson by the American Federation of Teachers. Therefore, while facing intense scrutiny from American educators, instead of reforming, Pearson decided to take its methods to poorer countries that are less able to mount a critical challenge to its claims. Desperate nations will come to rely on Pearson and its for profit schools over which Pearson has near autonomy over reporting itself a success or failure. In addition, because of these countries' already poor education, anything that Pearson does might look like success in the short-term.

All this information on Pearson Education, its operations, its employees, its leaders, and its general behavior made me and my attorneys more acquainted with what type of entity and people we were dealing with. It helped us begin to anticipate what lengths they may go to force me to do what it wanted. Because Pearson sent someone from its New York offices to Alabama, I realized that its executives did not consider this a small, local affair. While all this research into Pearson may not seem related to my case, I believe it is. If your legal opponent is going to go after your character, it's better that you do some oppositional research just in case you may need it. Besides, if Pearson was going to research me, I'd better research them.

Shady Law Firm

The intensity of Pearson's response was further highlighted by its hiring of Balch & Bingham, a large corporate law firm. Of course, you'd expect a billion dollar corporation to hire a high-powered law firm to represent its interests, but to me, something never felt right about Mr. Tankersley's and Balch & Bingham's antics. Independent of my attorneys, I thought I'd expand my research to include the law firm, and what I found was just as troubling as the information I learned about and found on Pearson.

Just as I had made the mistake of thinking that Pearson, being an education company, might behave reasonably and fairly, I was blindsided by Balch's actions. I expected more from a law firm that was well-established in the state, and I thought that they may behave ethically regarding the law and other areas of its operations. But as I stated earlier, Tankersley's aggression and threats toward me and towards my mom, made me think of them more as unscrupulous shakedown attorneys. Later events would demonstrate just how far Pearson and Balch were willing to go.

When Dr. Christopher Love, my former English teacher, and I started researching Balch & Bingham, it didn't take us long to discover that the law firm itself had earned notoriety. The law firm has over 200 attorneys and offices in Alabama, Florida, Georgia, Mississippi, and Washington, D.C. and boasts on its website of its "multi-disciplinary approach," such as representing "clients in the development of new nuclear facilities," two of the three largest banks in Alabama, and "the second largest investor-owned electric utility provider" (the Southern Company). The firm handles many other types of clients and law as well, including environmental, so it's clear that the firm is well-entrenched into the lucrative world of representing the wealthiest of clients. Despite its success, Balch has been involved in some ugly legal battles and scandals.

In 2018, Joel Gilbert, a former Balch & Bingham partner, was convicted of bribery, wire fraud, conspiracy and money laundering conspiracy. At the time of his crimes, Gilbert was part of Balch's environmental and natural resources practice and had bribed Alabama state representative Oliver Robinson with $360,000 on behalf of Balch client the Drummond Company. Prosecutors alleged that Drummond wanted Robinson to obstruct the EPA's expansion of a Superfund site, which is a polluted area that requires long-term cleanup efforts. The Drummond Company was among five industrial companies investigated by the EPA for responsibility for the pollution and contamination of the community of Tarrant, Alabama just north of Birmingham. Balch attorney

Steven McKinney was also on trial, but he was dismissed from the case based on problematic testimony by a prosecution witness. In addition to Gilbert, Drummond Vice-President Dave Roberson was also convicted while Oliver Robinson pleaded guilty to his charges.

In March 2019, Roberson filed a $50-million-dollar suit against Balch. Roberson claims that he never knew that Gilbert was breaking the law and was relying on Balch's legal advice during the EPA-Superfund affair, according to Roberson in an interview he gave to The Washington Post. Roberson also stated in the interview that Drummond Company, Balch & Bingham's client—not Roberson individually—had Roberson sign the checks to Balch & Bingham "to reimburse Balch for Robinson's grass-roots campaign" (referring to Robinson's attempt to interfere with the Superfund site) when "normally" Drummond's "general counsel signed checks." After Drummond fired Roberson, Roberson realized that something was up: he was fired the day after the statute of limitations had expired for suing Balch & Bingham for legal malpractice. Balch maintains that Gilbert acted on his own, but Roberson alleges that twenty-one Balch attorneys "played at least some part" in their involvement with Oliver Robinson, according to the Post. It also came out in the trial that Gilbert had exchanged emails with Balch partner Greg Butrus about the firm's dealings with Robinson.

More significantly, as it pertains to my case, a false pleading presented by Balch & Bingham in Roberson's civil case caught my eye, as brought to my attention by a CDLU-run reporting site, which has been featured in the Alabama Political Reporter for its "highly skilled research." The CDLU

(Consejo de Latinos Unidos) is a "nonprofit consumer advocacy group" founded by Hortensia Magaña and K.B. Forbes; on its main page, cdlu.org, it lists a Birmingham P.O. Box as a place to send donations, and it advertises itself as a national organization that specializes in healthcare charity. According to the Reporter, one of CDLU's sites is ban-balch.com, a site that reports on the alleged shady dealings of Balch & Bingham. The Reporter quotes a former federal investigator who explains that "Someone's spent a ton of money to compile this information" and that whoever is doing the reporting seems to have outstanding, exclusive inside information on Balch. The false pleading filed by Balch was an attempt to argue that the statute of limitations had passed and therefore Roberson's case should be dismissed, which ties back to Roberson's awareness that he had been fired the day after the statute of limitations expired. Thus, it seems that Balch was enacting its careful calculation and Drummond may have made it back when Roberson was let go. But, as the CDLU and Roberson point out, Roberson was not its client. According to court documents the CDLU obtained, in its false pleading filing, Balch omitted "a critical phrase in the law." For instance, Balch presented Judge Tamara Johnson a document that states, "Plaintiffs' claims against those attorneys—regardless of whether the claims are framed under the Alabama Legal Services Liability Act, or as a common-law negligence claim, or as 'escrow agent liability'—are claims that arose out of the provision of legal services by Alabama legal-services -providers." Balch ends its sentence there and continues "And in this State. . ."; however, Balch omits the phrase "to the plaintiff-clients of the Alabama legal-services-providers." Therefore, it should read:

". . . are claims that arose out of the provision of legal services by Alabama legal-services-providers to the plaintiff-clients of the Alabama legal-services-providers. And in this State . . ." The significance of this omission, the CDLU writes, is that it shows that Balch was cutting the phrase so that the judge would rule that the statute of limitations had passed and that the case should be dismissed on those grounds. CDLU writes that "Roberson's legal team provided Judge Johnson with a side-by-side comparison" to show the omission.

Overall, I find it strange that Gilbert and Roberson were singled-out individually in this statewide-famous case, and that Balch and Drummond evaded responsibility. Many others in Birmingham find it odd as well, including other journalists and courtroom observers. Moreover, the CDLU site and journalist Roger Shuler have chronicled other controversies and legal entanglements of Balch, including a 2018 FBI-investigation into illegal billing, accusations of racism in a Vincent, AL land-deal, the strange case of attorney Burt Newsome who suffered from harassment and legal troubles related to his battles and professional rivalry with Balch. Newsome's case is separate from the Superfund affair, but oddly enough, Newsome is now representing Roberson in his lawsuit against Balch. In 2013, Balch lawyer Clark A. Cooper was sued by the state of Alabama for his role in a corrupt scholarship fund. Cooper was allegedly fired from Balch in 2017, perhaps in relation to his role in the harassment of Burt Newsome, according to the CDLU.

In late 2018, Alabama Environmental Manager Scott Phillips and EPA Region 4 Administrator Trey Glenn were also indicted with charges related to the Superfund case. AL.com writer Ivana Hrynkiw reported in 2019 that Glenn and Phillips owned a company called Southeast Engineering and Consulting that was associated with Balch and Drummond and directly linked to the Superfund site and the Robinson scandal. Southeast Engineering was formed "to fight the EPA efforts to test and clean up soil in north Birmingham neighborhoods," according to the indictment, Hrynkiw reported. In a 2018 AL.com article, Kyle Whitmire wrote that Glenn and Phillips "worked with Balch to oppose the EPA." Balch's choice to become involved with Glenn is troubling in and of itself because Glenn has a history of run-ins with Alabama's state ethics laws, as detailed by AL.com's Dennis Pillion in a November 2018 article. One notable instance is that Glenn worked as a private consultant for a landfill company that imported "human sewage" from New York and New Jersey into proximity of Birmingham neighborhoods. The quotation and the more detailed story can be found in an AL.com article by John Archibald, who refers to the interstate transfer of waste into Alabama as "the poop train." As further documented by the site, Glenn and Phillips have worked in close circles for more than a decade, drawing sharp criticism and ire from those concerned about ethical and environmental violations and pollution in Alabama. Such histories could not have been lost on Balch & Bingham when the firm became bedfellows with Glenn and Phillips in its attempt to help Drummond block the EPA's attempt to expand the Superfund site, a place full of contaminants located near a largely African-American neighborhood. In fact, it's reasonable to suspect that it is precisely because of Glenn's history that Balch would find him an ally in combating strict oversight by the EPA.

Moreover, Principal Deputy Assistant Attorney General of the Environmental and Natural Resource Division of the Department of Justice is former Balch & Bingham (Washington, DC branch) partner Jeffrey Wood, who worked at the firm during the Superfund scandal. Wood joined the ENRD in January 2017 and had to be recused from the case. According to a May 2016 report by globenewswire.com, however, in 2016, while employed at Balch and during the Superfund case, Wood lobbied on Capitol Hill to discuss Superfund laws. Furthermore, Wood has been recused from so many cases involving Alabama energy and utility cases that former ENRD head John Cruden remarked in an April 2017 email to E&E News reporter Sean Reilly that Cruden had only been recused from one case during his tenure. Reilly reports that the "scope of Wood's recusals testifies both to the long reach of Southern Co., which operates as a power producer in Georgia, Alabama, Mississippi and Florida, as well as that of Balch & Bingham, which has more than 230 lawyers and lobbyists." Reilly further notes that "Among other customers, for example, is Luminant Generation Co. LLC, a subsidiary of Texas-based Vistra Energy that is fighting EPA's attempts to require new pollution controls on several older coal-fired power plants under the regional haze program."

The fallout of the Superfund site case continues to bleed over into other lawsuits and indictments and exposes some of the dirty dealings among Balch attorneys, politicians, federal and state regulators, and a charity used to funnel money (which Rep. Robinson was using to receive the bribes). Balch and its cohorts were doing this while trying to persuade the EPA to limit its expansion to identify if more land inhabited mostly by African-Americans might be hazardous, which could have or has had long-term health effects on the populace. Yes, Balch has an environmental legal department; but what and who it's protecting and whose interests they are looking out for remain suspect based on the Superfund case, the Vincent, AL land deal, and other cases and clients. Indeed, Michael Hansen, the executive director of an antipollution group in Birmingham (GASP) told Kelly Poe of AL.com that "if you need to fight an environmental regulation, you call Balch & Bingham."

The CDLU alerted me to this story; on its site, it posted a letter from Vincent resident Wanda Threatt to the Department of Justice. The story is supported and further detailed by Weld: Birmingham's Newspaper and journalist Cody Owens. The story involves a Florida company named White Rock Quarries, which was represented in Alabama by none other than, you guessed it, Balch & Bingham. According to Owens, Threatt's letter to the DOJ, and the CDLU's interview with Vincent residents, White Rock purchased land from white owners without offering to buy from black residents, leaving blacks with "unsaleable land" (qtd. from Threatt's letter). The CDLU reports that "local politicians, including a zoning official, allegedly sold their farmland for a nice bundle of cash, as if 'the fix' were in." Threatt wrote that "No offer at any price was made to a Black." The CDLU reported that these black residents insisted that "Balch & Bingham lawyers (and their public relations stooges) spearheaded the purchase of farm land"; this quotation and the quotation from Threatt's letter come from a Feb. 2018 CDLU article "Alleged Racism and Age-Discrimination in Vincent" (many of the black residents were senior citizens).

Residents expressed concern over the quarry because Vincent had already been the locale of an EPA Superfund site related to Alabama Plating Company, according to Owens. Furthermore, these residents pointed out that White Rock, despite what its president Jim Hurley claimed about White Rock's spotless record, was implicated in a water contamination scandal in Miami, FL and was ordered by a judge to stop its mining operations in the area. Several people also expressed concern about potential long-term cancer risks, and they did so from experience: they believe that the Superfund site from the plating company caused cancer in many of Vincent's citizens. In addition, environmentalists have expressed concern about the quarry's effect on the Coosa River and its groundwater. Cody Owens reported in 2013 that "White Rock's discharge valve is located only about a mile and a half from the intake valve for Jefferson County's drinking water." This reminded me of what I had read about White Rock's troubles in Miami when it was forced to halt operations because of pollution of residents' drinking water. As reported by Isaiah Thompson in the Miami New Times ("The Poisoned Well"), the pollutant in the Miami water was benzene, a carcinogen or cancer-causing agent. This is the out-of-state company that Balch assisted in coming to Alabama, despite the concerns of Alabamians who have seen their relatives suffer from cancers possibly related to previous pollution and environmental damage.

These concerns, though, did not deter White Rock, Balch, and the politicians of Vincent, who sided with White Rock in its legal battles against citizens. In one case, which made it all the way to the Alabama Supreme Court, White Rock and the Town of Vincent were represented successfully by Balch & Bingham partner Rob Fowler of the firm's "Environmental & Natural Resources Government Relations" division. The case allowed White Rock to proceed with the building of the quarry; however, more litigation has delayed construction. Despite Balch and White Rock's victory, Threatt's letter to the DOJ demonstrated that the fight had not ended. In a desperate plea to the US government, Threatt stated that "We face exposure to hazardous waste, air, water, noise and vibration pollution" (qtd. in Owens's 2013 article "A Big Hole in the Heart of Vincent"). Threatt even challenged town officials for ignoring the pleas of their constituents.

Owens, in the same article, also substantiates the CDLU's claim that a zoning official of Vincent benefited heavily from the sale of lands to White Rock. Robbie Greene, according to Owens, was a Vincent zoning official who sold her home and land to White Rock at approximately three times their worth, and she and her family members eventually pocketed at least $2.5 million while a black neighbor of hers received no offer from the company. Greene, notes Owens, was the sister-in-law of then-Vincent mayor Ray McCallister. Other white property owners also sold their land "quickly and quietly for highly inflated rates." The total amount of payout by White Rock to white residents totaled $11 million at the time of Owen's 2013 article. Subsequently, Balch represented White Rock in its fight against having land in Vincent designated as historical sites, and the case dragged on for several more years. In March of 2019, after winning several more legal battles, White Rock donated to the city of Vincent $600,000, the first of two payments in exchange for the cessation of further lawsuits to halt the quarry, according to Nathan Howell of *The Shelby County Reporter*.

Many facts and much evidence contained in these cases bothered me. In both the Superfund case and the Vincent land deal saga, the potential harmful effects of polluted lands mainly threatened primarily African Americans—and Balch & Bingham was on the side of the mining companies against them. Moreover, some of the people and entities that Balch represented or had close, mutually beneficial relationships with long histories, which were publicly documented, of abusing ethics, the law, and of having been negligent of environmental regulations or concerns of citizens, often black citizens. Based on phone calls and research on the racial makeup of Balch, I found that only about 2%, perhaps less, of Balch's attorneys are African-American. Considering that they operate in Mississippi, Alabama, and Georgia, states with black populations higher than 25% in addition to Washington, DC, a predominantly black city, these numbers show a dearth of black representation. I continued doing more research into the law firm and found other cases in which Balch represented large corporations against ordinary people and discovered that this history was plentiful as well. While it's true that Balch also has represented clients and cases in favor of environmental protection, I find Balch's environmental history disturbing. These other cases don't excuse its close relationship with companies whose actions and operations threaten the long-term health of black people and people in general in the state of Alabama. It doesn't excuse its lawyers from engaging in bribery, fraud, and other crimes coupled with many other ethically questionable decisions that have been documented by numerous sources, journalists, and investigators. Coupled with its pairing with Pearson Education in the case against me, Balch and its lawyers proved a dangerous adversary and one that no one would want to cross without significant legal representation.

So, what does all this have to do with me? One day, Dr. Love called me and told me something eerie.

"I was trying to do a little more digging," he said. "So, I made a few phone calls and wanted to have a closer look at some of their cases."

I didn't like the tone of his voice; I had no idea what he was about to say, but it didn't sound good. "Go on," I said.

"Well," he continued, "I made a phone call to request some of the case files involving Balch & Bingham. It doesn't matter whom I called; I think it's better the less I tell you in this instance. Let's just say I called whom I needed to."

I was getting a little nervous. I wasn't sure if he was just kidding, but that would have been unlike him to make light of my situation.

"And?" I asked.

"The person on the line was all willing to help me until I mentioned Balch & Bingham."

"Really?"

"Yeah," Dr. Love said. "As soon as I said Balch & Bingham, he became very evasive. Then, he started asking me questions. He actually asked me, 'Why do you want to know so much about Balch & Bingham?' I mean, it sounded like a line right out of a John Grisham novel."

I didn't know what to say. All I could do was just repeat, "Really?"

"I realized the guy wasn't going to give me any more information, so I just got off the phone."

I wondered just what I had gotten myself into. The patterns of behavior and tactics are troubling, and aspects of these patterns would appear as my criminal case moved closer to trial. Second, my research opened my eyes to the larger threat that I was facing and just how intricate the inner workings of a law firm and corporation like Pearson Education are. I realized how important it is to be aware of who your opponents are in business. This wasn't about competition; this was about an attempt to destroy my life. How does a college student combat the machinations of a powerful corporate law firm and a billion-dollar corporation that both have a history of playing loose with the law, ethics, and people's lives and who have powerful connections in government and possess a team of the best legal experts in the country? These were questions that haunted me as my case lingered, but one in particular kept coming to mind: What kind of law firm was I dealing with?

CHAPTER 8

Volte-Face

"They got you," my consulting attorney Daniel Fortune had told me. It was early in the summer of 2018. The words echoed in my ear. Because Mr. Fortune had been a federal prosecutor specializing in cybercrimes, his words packed a punch. He also had a no-nonsense demeanor, so any thoughts I had that he might be joking quickly dissipated. Indeed, I left his office that day when he decided that I didn't have much of a defense based on the evidence that the DA had presented. In the coming days, I would reach a point of despair. For two years, I had held my ground. I had dismissed any suggestion that I take a plea deal. I had finally got what I wanted: an attorney who had expertise in the areas of the indictment to hear my case, but he made it clear that I was guilty of at least some crime, if not all with which I was charged.

Inevitable Fate

Over the next few nights, I began to think about resigning myself to what seemed an inevitable fate. I had done all I could, and it didn't appear to have been enough. Although I tried to accept that I was going to be found guilty or might have to plead guilty to something to make it all go away, a part of me couldn't accept that everything that I had strived for was going to waste. I was going to suffer life-long consequences even after I had served a sentence or even if I just received probation. I was going to have a conviction and heavy restrictions placed on me that may prevent me from using a computer for a set amount of years.

Plus, the more I thought about it, the more I realized that even if I were to take a plea deal, the ending of my criminal case would not necessarily be the end of it. After I'd be convicted, Pearson, who would have tremendous leverage, could come after me in civil court seeking damages and make further demands. Furthermore, it would be possible for the University of Alabama to reexamine my status if I were to be convicted of a felony. Therefore, I could face expulsion.

Without a degree and with a felony on my record, my career prospects would be bleak. I could forget about a career working with computers for a long time to come. Moreover, with a felony conviction, I'd be lucky to get any job.

When I came to the University of Alabama, I had imagined a brighter future. With my skills and knowledge, I could work freelance, for myself, and change jobs at will if I wanted. I'd also be able to travel the world and, of course, be financially secure and make enough to help people, like my grandfather had done. I had already used some of my money to open the community center back in White Hall, and I'd been volunteering in Birmingham helping underprivileged, particularly African American, kids learn about computers. Many of these kids were shocked to learn that a kid who looked and talked like them could be interested in and know so much about computers. So, I'd been a potential role model for them because I showed them that it's okay for them to have an interest in something other than life in the streets. By seeing me and working with me, they could realize that they could learn a useful skill and that education was just as much for them as it is for anybody else. It was starting to hit home just how powerful the aftershocks of my case would be.

Furthermore, Attorney Fortune's grim conclusion exacerbated my personal problems. The case had already caused issues between my girlfriend and me; we were arguing more, and the case consumed my thoughts. Before the official indictment, I lived in fear that at any moment that the police could come and arrest me. I also had to wonder what Pearson was cooking up as it waited for the conclusion of the criminal case. My frustration festered, and I was caught between resignation and defiance. I counted sleep by the minutes rather than the hours.

Before returning from Google in 2017, I had wished for any type of movement on my case, but now that it was reaching its finality and not in my favor, I was rethinking my strategy and tactics over the past two years. I couldn't do anything about it, but those are the thoughts that haunt you. I wondered if I had just changed or said a few words differently if that would have made a difference. I wondered if I should have been more aggressive against Pearson. I didn't think, though, that I could have been more conciliatory because I had already told Pearson and its lawyers back in 2016 that I was willing to meet with them, an offer which they rebuffed because I wasn't going to meet their severe conditions. It's inevitable, though, to drive yourself crazy thinking about what could have been done differently when you start to recognize that breaks are not going to go your way, and these ideas and fantasies kept popping up in my mind which only added to my torment.

As the days passed, and I grew more anxious about my fate, I was, however, steadfast in at least one aspect: I was not going to take any plea that ended up with me going to jail or prison. That was final. If I had to take a plea, I might consider it, but I was not going to sign anything that put me behind bars. If I had to take my chances with a trial, then so be it. My resolution allowed me to still have pride in myself even if I had to accept a deal.

Then, a few days after our meeting, I got a call from Attorney Fortune. He told me that after thinking it over and reviewing the case a little more, he decided to represent me. Although he wasn't confident that I'd be cleared of everything and wasn't going to guarantee such total exoneration, I was ecstatic. I wasn't sure what had convinced him to take my case, but I was sure that I could answer all his questions and provide proof of my innocence.

I also had another person in my corner. In addition to Ms. Carter and Dr. Love, Professor Jeff Gray from the Department of Computer Science at the University of Alabama was willing to help. A full professor, Dr. Gray is a Distinguished Member of ACM, or Advancing Computing as a Science and Profession, a recipient of the National Science Foundation CAREER award, won Professor of the

Year from the Carnegie Foundation of for the Advancement of Teaching, and has received many other accolades throughout his career. He looked over my case and offered his opinion. After his review, Dr. Gray wondered why my case was even in the hands of a criminal prosecutor. He stated that as far as he was concerned that this seemed more like a civil matter rather than a criminal case. In an email to me and my attorneys, Dr. Gray wrote that "There was no intrusion into the Pearson websites" and supported my assertion that the information I obtained from Pearson was public: "anyone could have done this openly by the way their system was designed," he wrote. He also concluded that Pearson did not appear to protect personal and private information "very well by publicly exposing privacy details through publicly available API service calls." Thus, Dr. Gray confirmed what experts at Google and Cigital had to say about my case. If Mr. Fortune was still harboring any concerns, I knew Dr. Gray's findings could relieve them.

But Fortune understood that it's not enough for a defendant to convince his own attorneys that he's innocent; the people who need convincing are the prosecutors in the DA's office, or if it comes to it, the people sitting on a jury. The Tuscaloosa DA's office was already deeply invested in my case and had been for two years. After securing a grand jury indictment, the DA wasn't going to walk away even if we told them that we had a whole line of experts willing to testify on my behalf. I'm not accusing the DA of impropriety or corruption, to be clear. I believe, though, that the intricacy of my program and that technical complexities of the case made them rely heavily on evidence and rhetoric given by Pearson Education and its team. Pearson's manipulation of

UA officials using panic and outrage demonstrates how convincing could the corporation could be, and the statements given to UAPD by its software engineers proved compelling for the DA to surmise that a crime had been committed, and the prosecution was willing to let a jury make the decision about which team of experts they believed.

In April 2018, before Daniel Fortune came aboard, my attorneys David Schoen and Joel Sogol were still concerned about Pearson forensic analyst Shawn Garringer and software development manager Mark DeMichele's official statement to UAPD. To recap, Garringer had claimed that he and DeMichele had reverse-engineered my program. So, DeMichele and Garringer implied that they understood how my program worked and how I had created it. Consequently, his statement made it seem that he and Pearson could fully and accurately explain in court how my program was illegal. For instance, DeMichele had stated in the notarized police report that he determined my program was, as I earlier quoted, "disruptive and illegal." DeMichele's statement, taken in December 2017, closely followed the language that Mr. Tankersley had used as early as the summer of 2016 when he was emailing me and my attorneys. Tankersley had referred to my program as the "Cheating Program"; DeMichele used the terms "Cheating Software" in his statement to police a year later. To me, DeMichele's statement seemed rehearsed and carefully worded as if Tankersley or some other lawyer had coached him. Their words were too much in sync for me to think that these phrases were used coincidentally.

Additionally, Pearson and its lawyers were demanding that I hand over my source code, so if DeMichele had reverse-engineered my program they wouldn't need me to give them the code. I'm not sure what Pearson and Balch & Bingham thought they were doing regarding their claims. I suppose they didn't think I would read the reports and that my attorneys wouldn't be able to challenge the testimony of the Pearson software engineers. They were wrong. I sent Sogol and Schoen an email explaining how DeMichele's statement to UAPD sounded disingenuous regarding some of his claims. Regarding his echo of Tankersley's "cheating" claims, I stated on my tutorial site that the program should not be used for cheating. The criteria for cheating is to be established by the instructor or professor of whatever course a student was taking, so it would be up to the user to determine if they were following the guidelines set by the syllabus or by those laid out in another manner by the instructor. For example, as I stated before, my math teacher Ms. Carter had given me permission to use my program under certain conditions. Therefore, Tankersley and DeMichele's assertions that I had created a "Cheating Program" and "Cheating Software" were incorrect because they were not in a position to determine what constituted cheating for any class. Such a determination is up to the instructor or to be determined by a process within each university or school. When Tankersley stated in an email in 2016 that my program was a "Cheating Program," he apparently did not know or did not care that my own teacher had permitted me to use it. Second, when DeMichele reiterated the "cheating" characterization of my software in December 2016, he either did not know or did not care that the University of Alabama had cleared me of all academic misconduct.

Obfuscation & Encryption

Getting into the technical details, I explained to my attorneys that MathLabAnswers was a two-step process. A launcher allowed users to login and an application allowed users to obtain answers. The launcher controlled whether a user was able to download and run the application. I went back through my MathLabAnswers program source code when I realized that I had encrypted and heavily obfuscated the launcher. Encryption is the process of converting information or data into a code, especially to prevent unauthorized access; obfuscation attempts to make the encryption unclear or even unintelligible. The encryption algorithm I used was TripleDES (3DES) with a 192-bit (really 168-bit) key. This is the most secure form of this encryption algorithm.

From a reverse engineering standpoint, for someone to obtain the source code of MathLabAnswers they would have to do the following:

1. Bypass the encryption of the launcher by doing one of the following:
   1. Brute Forcing (randomly trying as many techniques or combinations as possible)
   2. Bypassing the obfuscation for the encryption key
2. Bypass the obfuscation of the launcher
3. Bypass the obfuscation of the application

To bypass the encryption, a reverse-engineer would have two options. Find the encryption key or bruteforce the encryption key. If brute forcing was attempted, the engineer would have to obtain a computer that can process 2^{192}

operations. To bruteforce a 128-bit encryption key would take 13.75 billion years with the best of today's technology. The key I used was 192, so it would take longer than the universe has even been in existence.

To bypass the obfuscation, a reverse-engineer would have one option: solve all the math produced by the obfuscator. An obfuscator is a program that can, for example, convert a simple line to print "hello" to a mathematical algorithm that can also print "hello." The original source code to the application was approximately 30,000 lines of code that I wrote. This does not include the external programs (common-lang library, guava library, Jxbrowser libraries, and Google-closure-compiler library) that I used to execute the application. So, in total more than 500,000 lines of code were converted to mathematical algorithms.

Pearson wrote up its 2016 report and sent it to the University of Alabama in under fifty days based on the timestamps in the report. It would have been impossible for anyone to reverse-engineer MathLabAnswers in that time. Thus, I told my attorneys that they needed to ask the DA how Pearson could claim that its software engineers had decrypted my software. Another point I wanted my attorneys to remember, if they hadn't, was that Pearson had told the UAPD that its chief concern was the exposure of confidential student information that Pearson was obligated to protect; this concern was what started the whole fiasco. Pearson's subsequent claims grew out of its horror that I had revealed that the corporation had left millions of students' personal information unprotected and publicly available. From that initial point, the evidence shows that Pearson was trying to bend the narrative that I had done something illegal, and after my arrest was secured by Pearson through the UAPD, its employees and lawyers no longer wanted the discussion of its security lapses and vulnerabilities to be the topic of discussion. Such a revelation could be devastating to Pearson financially, leading to schools withdrawing or canceling contracts or letting them expire. Therefore, Pearson was in full survival mode. If it needed to fudge on or lie about the details, then it would. The fact is Pearson's employees had stated and implied to the police in an official document that they had reverse-engineered my program when they had not. Because they had lied about that, there was no telling what else they would lie about. Lo and behold, this would not be Pearson's final lie.

Logical Fallacies

Because I had caught Pearson's employees' deceit, I thought for a moment that this would be a turning point in my case. After DA Sarah Folse realized that Pearson had not reverse-engineered my program, I believed she would see that I had exposed a major problem with the state's evidence and that her case was beginning to fall apart. To reiterate, because Pearson did not reverse-engineer my program, its employees and certainly its lawyers had no idea how my program worked. Not only had Mr. Tankersley demanded my source code, he had also wanted me to explain how the program functioned. Therefore, he could not know if any law had been broken. It seemed to me that the entire case hinged on whether Pearson's engineers and the DA could provide honest testimony from any expert who could state in detail and step by-step how my program worked and how I went about creating it to prove to a jury that I had broken the law. If the DA planned on using Pearson's employees, I knew that I could easily debunk their claims of reverse-engineering. Furthermore, if Pearson was going to argue that my reverse-engineering of its programs was illegal, then I could argue that its employees' claim to have reverse-engineered my program would make their actions illegal as well. I believed that I had pinned Pearson into a legal corner and that it was tying itself into knots trying to argue its case against me.

DA Folse, though, would not be dismayed by Pearson's attempt to sneak misleading terminology past investigators. She interpreted Pearson's language as a technical error, a trivial mistake that did nothing to dissuade her intent to pursue prosecution. The DA's reaction stunned me because I thought that by now Ms. Folse saw the weakness of the case and that Pearson had misled her office. More importantly, I had hoped that exposing Pearson's problematic testimony and all the subsequent questions and points my attorneys raised would begin to make the DA realize that significant grounds for reasonable doubt existed and that I hadn't broken a law and that at best, as Dr. Gray had suggested, Pearson only had a feeble civil case. It was hard for me to accept that Ms. Folse believed that I was 100% guilty of all the crimes on which I was indicted. It's possible that Ms. Folse may not have been making the final decision about the prosecution of the case. Because Pearson had already influenced the University of Alabama's administration to come after me, I reasoned that it was possible that Pearson may have pressured other officials in the District Attorney's office to prosecute the case. Second, based on my research, I knew that Balch & Bingham had powerful political connections throughout the state. Surely, Balch & Bingham attorneys were familiar with Tuscaloosa prosecutors. Was it possible that they could influence the direction of the case? After all, during the events of my case, one of their attorneys had been on trial and convicted of bribing a state representative to advance the interests of his client. These possibilities crossed my mind, and I considered that even knowing what I knew about Pearson and Balch I was still underestimating their power and reach. I had to keep reminding myself how much was at stake for Pearson. The corporation needed my conviction to fortify itself against potential claims and fallout for leaving students' confidential information publicly accessible. Again, we're talking about tens of millions, if not more, in possible losses.

It's not that the DA's office necessarily had to be corrupt. It could be, I considered, that it was susceptible to overstatements and deception by Pearson and Balch. Whatever it was that made DA Folse obstinate, it certainly wasn't the evidence that she had on hand. Regardless of her reasoning, I had to come to terms with the case's inertia. Once again, I had to rethink my dwindling options: plea or go to trial.

When Attorney Sogol had emailed me about my charges while I was at Google, he had mentioned that we might be able to petition for youthful offender status. In Alabama, "Youthful Offender" status can be applied to people under twenty-one who have been charged with a crime. If the charge is a felony and the court grants a defendant YO status, then the defendant will face a significantly reduced sentence and have the conviction, charge, arrest, and other court actions sealed. As the case continued to head toward trial, I had to wonder if I was going to have to consider such a deal or any other offers that the DA might proffer if it didn't involve any jail time.

Then it happened. Pearson Education and Balch & Bingham couldn't help themselves. I suppose these two bedfellows had been pulling stunts for so long that it was second nature to them, just another tactic to employ. And it didn't matter if it was illegal, immoral, or unethical. In their world, it was just another day at the office.

It's possible that these two companies were getting antsy and frustrated with the pace of my case. I had thought that they were pleased with how the investigation was moving along. A grand jury indictment had been secured, and though the case was dragging on from my perspective, I didn't know what Pearson thought about its pace. The corporation was no doubt used to drawing out legal matters, and because Balch was getting paid, its partners probably didn't mind cashing more of its client's checks. But Pearson may have been worried that the DA and my attorneys were going to negotiate too leniently of a deal or word may have gotten back that I was willing to go to trial, which could expose to the public Pearson's security lapses and negligence. Thus, the billion-dollar corporation might have pressured its lawyers to find a way to end this case as quickly and quietly as possible. Or these two cohorts may have had this tactic in their bag of tricks all along, waiting for the opportune moment to spring it.

One of the questions my legal team and, on her end, the prosecutor had was about the nature of the End-User License Agreement, or the EULA, that each student had to electronically sign when using Pearson's MyLab products. Pearson had supplied the DA and my legal team with a copy of the EULA to prove that I had violated the contract. My legal team and I, though, questioned whether violating a EULA with a private company constituted evidence of criminal behavior. The dispute over a EULA agreement seemed more like a civil dispute to be resolved in civil court rather than something that should warrant or be a basis for a criminal charge. For my supposed violation of the EULA, Pearson had already disabled my Pearson MyLab account back in 2016. Usually, disabling of their accounts is the only penalty for such a violation that users incur.

The strategy, though, was clear. Pearson and Balch could accomplish two goals. They could use my alleged violation of the EULA as evidence that I was up to a criminal act, even if violating a EULA may not be considered a crime. Violating the EULA could provide the context through which a crime was committed. Furthermore, they could get the prosecutor to declare in open criminal court or have it documented in court records that I had violated the EULA. Consequently, when they sued me in civil court, they could use the documentation from the criminal case to substantiate its civil claims.

In July 2018, as the trial approached, DA Sarah Folse sent Balch & Bingham attorney Michael Taunton an email about the EULA agreement. Mr. Taunton, according to his resume posted on Balch's website, focuses on business litigation and had "served as legislative counsel to Senator Jeff Sessions" and "specialized in matters of intellectual property." Taunton and Tankersley practice in the same areas of the law, so based on their years of experience I presumed that Taunton was assisting Mr. Tankersley with the case.

Taunton, or another attorney from Balch, sent earlier a copy of the EULA agreement to Ms. Folse, so it's clear that Ms. Folse had been in communication with Balch's attorneys. Such communication heightened my suspicion that Balch may have had considerable influence over the course of the criminal case. Ms. Folse requested in her email to Taunton that he provide her with the EULA agreement that was in effect at the time of my 2015-2016 access to Pearson's MyLab software. Taunton replied with a copy of the EULA, which was forwarded to my attorney Joel Sogol from Ms. Folse.

A EULA is almost always long, of course, and most people don't waste any time reading it. In fact, many companies take advantage of their lengthy EULAs to include conditions that users wouldn't agree to if they had the time to read the agreement or if the service wasn't part of a service that they had to use. Most people are willing to trust that the terms of a EULA are more or less fair for the sake of the convenience or necessity of using a product. A person may not want to agree with everything listed in their cell phone EULA or when using an app, but they're willing to overlook their concerns just to use the service. Thus, companies know that they're not going to have trouble getting users to agree to their terms. Moreover, I find it strange that a EULA can be forced upon students who are required to use a company's software for their classes. Students are compelled to "agree" to a company's terms because they have no other choice; otherwise, if they didn't agree to the terms, then they wouldn't be able to complete the assignments or tests, and they would fail the class. This allows "education" companies like Pearson to make the most of their EULAs. This is what Pearson did with its EULA in its Project MASH—

which I discussed earlier—when it was able to compel users to acknowledge that their personal information could be shared and used for future Pearson purposes. College students aren't making consensual choices; if they don't want to use the product, their only option is to not take the course, which is not an option regarding courses that students are required to take. Plus, Pearson's EULA is almost 8,000 words (about thirty pages by conventional measures). Because they do not have a realistic option to opt out of using a product like Pearson's MyLab, students have even less of a reason to read its lengthy EULA.

But if Pearson and Balch & Bingham thought that I wasn't going to or had not read the EULA, they were mistaken. I have experience working at Cigital/Synopsys, Enjin, and Google, so I know the language, conditions, and criteria of general EULAs like the one Pearson had for its MyLab programs. Although Pearson and its lawyers were apparently counting on the EULA as its trump card, I had already taken into consideration its terms and conditions when I began to create my program back in late 2015. In fact, these were the reasons I all along felt so confident that I had not broken the law. Indeed, one of the questions I asked my attorneys is whether the EULA even applied to me because when I explored Pearson's site, I was not under my student account but its general, publicly accessible demo site. Before we could determine the answer, though, we still needed to review the EULA.

Forging Documents

In his reply to Ms. Folse, Balch attorney Michael Taunton sent a copy of the Pearson EULA and instructed her to pay attention to Sections 3 and 6. Section 3 has to do with the use of Pearson's website information and materials; Section 6 has to do with the further restrictions on the use of the website and what the user agrees not to do. More specifically, Mr. Taunton told Ms. Folse to read especially Sections 3.3 and 6.2-6.7 and 6.12. Ms. Folse forwarded Taunton's reply to Attorney Sogol, and together my attorneys and I reviewed the EULA, reading it line-by-line as carefully and closely as we could.

It didn't take long for my team and I to notice something was fishy. Attorney Sogol spotted it right away and noted his suspicions to me in an email on July 16, 2018. I forwarded the email to David Schoen. I began to search for the EULA agreement that I would have been bound to when I created and ran my program. Apart from the EULA that Taunton had copied into his email to Ms. Folse, I found a different EULA agreement that was purportedly from 2016. Attorney Sogol noted that on this version of Pearson's EULA agreement it stated that it had been "Revised June 23, 2017." My program and update were created in 2015 and 2016, respectively; the program was shut down in its entirety by June 2016, so the revision date raised our eyebrows. If Pearson had revised the terms of its EULA, even in part, in response to my program, this meant that I was not retroactively responsible for conditions it added later. Also, Pearson had another version of a EULA that was longer and had been in effect since 1997 and had last been updated in August 2016—once again, after my program had already been shut down. So, we had been trying to confirm with Ms. Folse which EULA Pearson was claiming that I was under during December 2015 to June 2016.

Our request is what prompted Ms. Folse to email Michael Taunton to clarify which EULA Pearson claimed governed the period I was a registered user for its MyLab site. When Ms. Folse forwarded Taunton's email with the EULA that he asserted was the correct one, we began to compare that EULA to the language to the others that we found. And we found major discrepancies and suspicious language that seemed all-too convenient for Pearson and Balch & Bingham.

We're not sure when, but based on dates, Pearson and Balch had supplied the DA after June 2017 with a copy of the EULA that they claimed I had signed back in 2015 and was in effect to the spring of 2016. In preparation for my trial, Attorney Sogol had also obtained a copy of this EULA. This EULA was the EULA that was marked "Revised June 23, 2017." Balch & Bingham attorney Michael Taunton had highlighted or authorized the highlighting of several passages from this EULA. This highlighted EULA contained a new language that was not in the EULA that I had signed back in 2015. When Attorney Sogol made Ms. Folse aware of this discrepancy, this is when she emailed Michael Taunton for the correct EULA—the one that I had signed. As late as July 2018, Taunton and Balch were trying to pass this EULA off as the one I had signed. Based on the record from the electronic document, Taunton had made his highlights of certain passages, ones he tried to use to show that I had violated the agreement, on July 11. This ruse, however, did not work. Instead, Ms. Folse, acting on Sogol's concerns, replied to Mr. Taunton on July 13 that "the user agreement you sent me has the date July 11, 2018." Thus, Ms. Folse had at least been convinced by Sogol that she had been given the wrong EULA by Balch & Bingham. If providing the wrong EULA were a mistake, it was a mistake that Balch wasn't going to correct. In fact, Michael Taunton was going to double-down on it.

When Taunton replied to Ms. Folse's request for the correct EULA, he provided a new one and stated that "Below is the EULA that was in effect when Desmond Jackson created his account with Pearson Education (and through August 2016...)." This EULA was indeed different from the one he had provided her, but oddly, the EULA he sent was pasted into the email. He added that, "We are also working to provide this to you in a documentary form." The email seemed strange. How long would it take to put a EULA taken from Pearson's website into document form or have Pearson send them the document? A few minutes? Why would Balch & Bingham need to "work" on putting it in document form as if this were a lengthy process? Plus, the highlighted EULA had been in document form. Why was the one he pasted into the email not already in document form? Ms. Folse forwarded the email to Attorney Sogol. After reviewing the pasted "EULA," Sogol forwarded the email to me and remarked that he was "convinced that they are pulling it off another website and trying to make it look like the one you used. What B.S."

One reason that Mr. Taunton may have pasted the new EULA was because in the highlighted EULA, the one in pdf document form, revealed the dates that the highlights were inserted, which was July 11, 2018. This means that by copying and pasting the new EULA the dates of any changes made to an original document would not appear.

Pearson and Balch & Bingham were in a bind. DA Sarah Folse knew that the first EULA she had been given by Balch, the one Taunton highlighted, was not the user agreement that I signed. Pearson and Balch knew it, too. They knew what language the original EULA that I had signed said, and they knew that it would blow a catastrophic hole in a potential civil case against me and the criminal case—the latter of which they had pinned their hopes on to cover themselves from civil litigation from users whose information they had left exposed, which numbered in the millions. And they also had to cover themselves from losing millions upon millions from contracts Pearson had with schools around the world that might sever ties because of its public exposure of student records. What could Pearson and Balch & Bingham do? Enter Michael Taunton.

Balch & Bingham had to figure a solution fast. My criminal trial was going to begin either just before or after the beginning of August, and it was already the middle of July. When Ms. Folse emailed Taunton on July 13, Taunton must have decided that he had to move. I don't know if he acted with the blessing of Mr. Tankersley but judging from the evidence I've gathered about Balch & Bingham and Pearson Education, I find it hard to believe that their entire team wasn't in on it.

Taunton had highlighted the first EULA sent to the DA to show the specific areas of the agreement that he claimed I had violated. When he sent the new EULA to Ms. Folse, he also had directed her to specific sections there as well. If you recall, these were sections 3.3 and 6.2-6.7 and 6.12. When we were reading this latest version of the EULA, Sogol, I, and my girlfriend, noticed a problem.

"Something doesn't look right," my girlfriend told me. The copy and paste job, the language of the email, and, of course, the history of their antics told us all that Balch & Bingham's lawyers were trying to pull a fast one. If my freedom hadn't been at stake in this case, their antics would've been downright comical if not pathetic. You wouldn't think that a multi-million-dollar law firm with connections all the way up the political chain to the US Senate would resort to chicanery, but then you probably haven't met the classless shakedown attorneys from Balch & Bingham.

After doing some searching, I found a copy of the EULA that I had signed, the one that was in effect for all users from 2011 to as late as August 2016, a period that covered all the events of my case. Along with Sogol, I read the entire agreement and compared the real EULA with the EULA that Taunton had pasted into his email to Ms. Folse. I looked closely at the sections to which Taunton had pointed. In Section 3.3.2 of the agreement that I signed, the real EULA, reads, "You may not post Materials from the Website nor answers to any homework or test questions provided by the Website to newsgroups, mail lists, electronic bulletin boards, homework sites, content aggregators, file storage services or any other on-line destination." This is the end of this section. In Taunton's version, however, more wording is added. It reads, "You may not post Materials from the Website to newsgroups, mail lists, electronic bulletin boards, homework sites, content aggregators, file storage services or any other on-line destination. Except as necessary in connection with performing the assignments contained in the Website, You agree that any posting of questions or other assessment content contained in the Materials and/or solutions to such questions or assessments (whether generated by third party or otherwise provided in the Materials) will not be considered 'fair use' and is prohibited under these this Agreement." I found another example of altered wording in section 6, the other section to which Taunton had directed Ms. Folse. In the EULA I signed, section 6 differs drastically from the section 6 that Taunton included in his email. In fact, many parts are drastically cut from the original and other sections are totally rewritten and wording is added to target the specific actions that Pearson believed I had taken in creating my program. For example, here is an excerpt from section 6 of the EULA that I signed:

1. Pearson does not guarantee the accuracy, integrity or quality of End User Content and disclaims all liability for any errors, omissions, violation of third party rights or illegal conduct arising from End User Content posted, emailed or otherwise transmitted via the Website.
2. Prohibited End User Content. You agree not to upload, post, email or otherwise transmit any End User Content that:
    1. is unlawful, harmful, threatening, abusive, harassing, tortious, defamatory, indecent, offensive, vulgar, obscene, libelous, invasive of another's privacy, hateful, or racially, ethnically or otherwise objectionable;
    2. You do not have a right to transmit under any law or under contractual or fiduciary relationship (such as inside information, proprietary and confidential information learned or disclosed as

part of employment relationships or under nondisclosure agreements);

3. infringes any patent, trademark, trade secret, copyright or other proprietary rights of any party;
4. contains software viruses or other contaminating or destructive devices, features or any other computer code, files or programs designed to interrupt, destroy or limit the functionality of any computer software or hardware or telecommunications equipment or provides information relating to or otherwise facilitating the use of malware or other destructive materials or mechanisms;
5. contains advertising or promotional material of any kind;
6. contains a third party's Personal Information (as defined in Section 8.2) except in connection with the posting and/or transmittal of grades or other information by an instructor or educational institution in a manner which may be accessed only by the relevant end user and/or which is consistent with Section 8.4;
7. undermines the pedagogical purpose of the Material on the Website and the course with which it is used, including but not limited to answers to questions used in the Website).

3. Pearson's Use of End User Content. By posting End User Content, You grant Pearson a perpetual license to host, transmit, use and distribute such End User Content for the purposes for which it is posted. You acknowledge and

agree that Pearson reserves the right to preserve End User Content: (a) for the longer of either the end of the relevant semester or other registration period for the relevant course or other website or for as long as the end user maintains an account with Pearson, and for a reasonable time thereafter in either event; and (b) indefinitely in the event of a potential violation of these terms or dispute pertaining to End User Content, the Website and/or the end user's account. You acknowledge and agree that Pearson further reserves the right to disclose End User Content in association with Your Personal Information if required to do so by law or based on the good faith belief that such preservation or disclosure is reasonably necessary to: (a) comply with legal process; (b) enforce the terms of service hereunder; (c) respond to claims that any End User Content violates the rights of third-parties; or (d) protect the rights, property, or personal safety of Pearson, its users and the public. In the case of End User Content embodied in Your performance of educational assignments and assessments, Pearson further reserves the right to host, transmit, use and distribute such content without reference to Your Personal Information indefinitely for the limited purposes of product development and educational research.

The section "6.4" statement is the end of section 6 in the EULA I signed. It then moves on to section 7. However, in Taunton's email, not only are sections 6.2-6.4 radically different from the one I signed, he adds new sections: sections 6.5-6.12. These are Taunton's changes and additions:

1. impersonate any person or entity, or falsely state or otherwise misrepresent Your affiliation with a person or entity; including using another person's Login Credentials
2. use or attempt to use any "deep-link," "scraper," "robot," "bot," "spider," "data mining," "computer code" or any other automated device, program, tool, algorithm, process or methodology or manual process having similar processes or functionality, to access, acquire, copy, or monitor any portion of the Website, any data or content found on or accessed through the Website or any other Pearson Materials without the prior express written consent of Pearson;
3. obtain or attempt to obtain through any means any Materials, End User Content, or any other data, content, software or code available on or through the Website ("Website Content") that have not been intentionally made available to You either by their visible display on the Website or their accessibility by a visible link on the Website
4. violate any measure employed to limit or prevent Your access to the Website or Website Content;
5. violate the security of the Website or attempt to gain unauthorized access to the Website, Website Content, or computer systems or networks connected to any service of the Website through hacking, password mining or any other means;
6. interfere or attempt to interfere with the proper working of the Website or any activities conducted on or through the Website, including accessing any Website Content prior to the time that it is intended to be available to the public on the Website;

7. take or attempt any action that, in the sole discretion of Pearson, imposes or may impose an unreasonable or disproportionately large load or burden on the Website or the infrastructure of the Website;
8. disrupt the normal flow of dialogue or otherwise act in a manner that negatively affects other users' ability to engage in real-time exchanges or to normally post messages, articles, or submissions;
9. interfere with or disrupt the Website or servers or networks connected to the Website, or disobey any requirements, procedures, policies or regulations of networks connected to the Website;
10. violate any applicable local, state, national or international law or the academic rules or other policies of Your Sponsoring Institution; or
11. engage in any conduct which otherwise diminishes the pedagogical or commercial value of the Materials.

Because Pearson's employees didn't know how I had created my program, meaning they hadn't reverse-engineered it, the EULA that Taunton sent to Ms. Folse had to prohibit acts that Pearson thought I had committed in the creation of MathLabAnswers. For example, in Taunton's 6.2 condition, which is completely different than the original 6.2 condition that I was under, it states that a user cannot "impersonate any person or entity, or falsely state or otherwise misrepresent Your affiliation with a person or entity; including using another person's Login Credentials." Although I had not impersonated anyone or used anyone else's login credentials or violated any of these conditions, these conditions were most likely written because Pearson may have believed that

I stole credentials; therefore, Balch knew that they had to cover these actions in the concocted EULA. Another example of this type of fishing for what I had done is the use of the prohibition against "password mining": the act of trying to obtain illicitly a password, which I had not done either.

Other conditions that are listed mention actions and techniques that are not illegal and that I had performed, like creating or using code and obtaining data (which is vague) that is connected to their site but were not prohibited in the original EULA that I had signed and to which I was bound. In short, Taunton and Balch & Bingham sent the Assistant District Attorney of Tuscaloosa County a fake EULA agreement to try to ensure my criminal conviction. They didn't care that their fake document could send me to prison.

There it was. Even after all I had been through—the police raid, the threats to me and my mom, the lies, digging into my past—I still had a hard time believing that a law firm, full of lawyers boasting on their biographical pages about their prestigious degrees, awards, acclamations, talents, expertise, and even adding little touches about how much they care about people and causes—had forged a document against a twenty-year-old college student that had the potential to destroy his freedom and his future just to cover for its billion-dollar corporate client. It shouldn't have surprised me. I knew that stuff like this happened more than anyone would like to think. People get railroaded into prison all the time, every day, everywhere. Law firms representing big corporations against individuals have a long history of wrecking lives, bankrupting their opponents, dragging out cases for years to break them, influencing criminal proceedings that could benefit their civil cases, committing countless unethical acts, and even straying themselves into criminality. Balch & Bingham certainly knew about criminality—after all, one of their own got caught bribing a state representative. I knew this. My grandfather had faced corruption from supposedly upstanding officials and lawyers his entire life. He needed good lawyers like mine—Sogol, Schoen, and Fortune—to protect himself. He's seen corrupt and racist politicians, corrupt and racist police, and corrupt and racist courts. I knew all this was out there and was happening and that even some of this corruption was happening to me, but I still had a hard time believing that these thugs had gone this far and faked a document and given it to a prosecutor. It didn't take me long to get over my shock because I had long known what kind of lawyers they were. I was no longer surprised that they had pulled this stunt, but I was still surprised by the sheer idiocy of the way they did it. I thought they could've come up with more sophisticated corruption then pasting such an obvious forgery into an email. I had a hard time accepting that they believed my lawyers were that gullible. Mr. Sogol is a well-known lawyer in Tuscaloosa and if you're a well-known lawyer in Tuscaloosa you're a well-known lawyer in Alabama. David Schoen has practiced law for decades and is based in Montgomery, so he's no stranger to Balch. Daniel Fortune was a federal prosecutor for North Alabama and was involved in the Superfund case in which the Balch lawyer was charged with bribery. Plus, Taunton and Tankersley should have known that a copy of the original EULA that I had signed was still accessible. But I suppose the hubris of Balch & Bingham knows no bounds. Or maybe in their desperation to save their case and client they decided to roll the dice and try to sneak one past everybody, hoping we wouldn't read the EULA.

You might think there might be some repercussions for Balch & Bingham, Taunton, and Tankersley, but there were none. In fact, even in light of the faked document, DA Sarah Folse was virtually unmoved. Though at some point she admitted to my attorneys that this whole case sounded like a civil matter, Ms. Folse was still not willing to drop all the charges. Instead, she began to offer plea deals to see if I might take the bait. I can only guess at her motivations, but it seems logical that the DA's office felt that it had too much time and work invested in the case and that it needed to save face. The DA might have also believed that I would feel relieved if I was offered what she considered to be generous terms. Indeed, I was offered a deal in which I would avoid jail time, be granted Youthful Offender status, and I would have to stay away from computers for two years.

Though I had a chance to escape jail time, after what Balch & Bingham tried to pull I wasn't in the mood for accepting a plea. I was also pissed off that the DA still wanted to pin something on me. And the DA expected me to not use a computer for two years? Hell, no. To be honest, the times when I considered taking a plea, I had thought about that possibility, that a deal might involve me not using a computer or accessing the internet. I thought that I might become a truck driver or something like that, but the more I thought about what all Pearson and Balch & Bingham had done, the more I resolved to continue to fight for my exoneration.

At this stage, Daniel Fortune became more involved in my case, and after he decided to take it on he began to relay the plea offers to me. When we initially met, he had tested me, telling me, if you recall, that it sounded like they had "got" me. A few days later, when Attorney Fortune decided to take the case, we began to put together a strategy for my defense. Mr. Fortune was so disconcerted with what had happened to me over the previous two years that he agreed to take on the case at a considerable discount. The DA's decision to press onward with charges left me and Fortune incredulous, but we were confident that if it went to court that we could offer the expert testimony to rebut the prosecution's claims—which we knew were all based on what Pearson had told them—we couldn't wait to bring up the conduct of both Pearson and Balch & Bingham. No one wants to go to trial, though, if they don't have to, this includes many prosecutors. They'd rather get the plea deal, have the conviction on record, and move on to the next case, like a coach piling up as many easy wins before they have to play a big game. So, DA Folse tried to eke out the victory in the final minutes.

That's when Mr. Fortune made it clear to her that if this case went to trial, she wasn't going to win. The Tuscaloosa County District Attorney's office isn't exactly stacked with an all-star team in the area of cybercrime. The kinds of crimes committed in Tuscaloosa don't usually involve a computer unless it's ripped from somebody's house. In other words, central Alabama isn't quite Silicon Valley.

As the eve of trial approached, Mr. Fortune served one final warning to Ms. Folse. If she insisted on going to trial, she should keep in mind at least one thing and be aware of another. She should keep in mind that he served as a federal prosecutor with cybercrime as his specialty. In short, he was the US government's all-star cybercrime prosecutor in the state of Alabama. And she should be aware that the law professor who taught her about cybercrime laws, well, that professor learned it from him; Daniel Fortune pretty much wrote the curriculum.

Dismissal Agreement

Suddenly, there was a dramatic change. At some point, Ms. Folse contacted Balch & Bingham, and she offered new terms to Fortune. The DA was willing to drop all charges if I met three conditions that Folse had negotiated with Balch & Bingham. Not only would my criminal charges be dropped, but Pearson would agree not to pursue any civil litigation against me. I've summarized the conditions here:

1. I would sit in person with Pearson's attorneys for a deposition and answer questions about the creation of my program, how the program engaged with Pearson's software, and questions about the information I had learned about the nature of Pearson's products. My answers, though, could not be "used as a basis to find [me] guilty of any crime," and should not "be interpreted or understood to waive" my rights and should not "be used as a basis for any civil liability."
2. I would turn over or destroy all Pearson data and turn over to Pearson and the state of Alabama all software I created that interacted with Pearson products
3. I agree not to violate any of Pearson's EULA agreements associated with its products and never again create software that interacts directly or indirectly

with Pearson's products or "undermines" their "pedagogical purposes" (Basically, don't do this shit again!).

Believe it or not, when offered this deal I was still hesitant. In just a few short months, from April 2018 to July 2018, Pearson's and the DA's cases had completely fallen apart. Simple discovery had exposed the corrupt nature of Pearson and Balch & Bingham, and none of its key players had even been questioned by my attorneys. I could only imagine what Daniel Fortune would have done to these jokers on cross-examination. Did Pearson want to have to defend the forged EULA? Did Balch & Bingham? Wouldn't they be subject to a criminal investigation in the aftermath? Did Pearson want me to testify about how its programs exposed millions of students' private information via a publicly accessible website?

As far as I was concerned, the tables had turned. I should be the one calling the shots. Pearson and its lawyers should be at my mercy. Sit for a deposition? During which Balch got to grill me over what I had created? I'd be helping Pearson and its software engineers, teaching them, and explaining to them the vulnerabilities of the software they created. Hell, I should get to charge them. After all, I'd be serving as their unpaid cybersecurity consultant. Fuck no because fuck them. If anything, by not going to trial, I'd be saving their asses from perjury and other crimes. Shit, when the criminal case was over, I was planning to sue Pearson.

In fact, when Fortune told me of the offer and sent me the terms in writing, I decided to consult with David Schoen. Because Schoen was on my team to handle the potential civil litigation, he was monitoring my criminal case to prepare to defend against any civil action Pearson might take against me. Schoen and my grandfather had known each other for years, dating back to my grandfather's days as mayor of Whitehall. After we consulted, Schoen had some reservations about the DA's offer to drop all charges if I sat with Pearson. On August 2, 2018, Schoen wrote to Fortune and Sogol expressing our concerns. His first concern was the condition that I be honest and complete in my statements to Pearson and its lawyers. Schoen was disgusted with Pearson's and Balch & Bingham's behavior and didn't trust them not to pull something in coordination with the DA. He wrote that "My fear is that they (Pearson, the government or both) as the decider of whether he was honest or complete." Schoen feared that Pearson might be setting me up for a perjury trap. Or, at the least, to claim I had breached the agreement and come after me in civil court. Schoen was willing to defer to Fortune and Sogol's judgment because they were handling the criminal portion of my case, but he just wanted to make sure that all possibilities were considered before I signed my name to an agreement with parties that couldn't be trusted. Fortune, though, replied that he had considered just such a possibility, but he was confident that I could be easily defended from a perjury charge if the DA stooped to such a low. More importantly, he advised Schoen, it was better that all criminal charges be dropped and if Pearson came after me in a civil court, civil court was a better arena than criminal court. Sogol concurred with Fortune, so I had to consider my options based on this discussion.

Part of me was still defiant to do anything under Pearson's terms. Part of me wanted to go to criminal court rather than take the deal so that Pearson's employees and maybe even Balch's attorneys would have to resort to perjury to try and get me convicted—and then we could really turn the tables on these folks. I thought that a criminal trial could shine the light on all Pearson's and Balch's lies and corruption.

But after contemplating my situation, I knew I had to let cooler heads prevail. The world doesn't work the way you want it to, so most of the time you just have to take the best deal you can manage. If you go to trial, there's no telling what could happen. In my request for revenge, I could wind up in prison. Who wants to sit in prison knowing that at one point you could have walked away scot-free? Believe me, there's nothing more that I wanted at that point in my life to see Pearson and the lawyers of Balch & Bingham get what was coming to them, but when I thought about the situation I had to realize that it wasn't likely that my fantasies were going to come true. Even though I believed that Pearson and Balch had broken the law, it didn't matter what I believed. If the DA's response to Michael Taunton's forged EULA was to still pursue my prosecution, then it wasn't likely that holding Balch or Pearson accountable for their actions was anywhere near the thinking of the District Attorney's office. Besides, by taking the offer, I could still sue them. The agreement terms did not prohibit me from pursuing civil litigation. I thought that was a strange omission on Pearson and its lawyers' part because they had to know that I could sue them. You'd think a billion-dollar-corporation would have had attorneys who would have anticipated such a move. But they didn't. I was also surprised that Balch did not demand a gag-agreement. But that wasn't in there either. So here I am talking about it—freely.

11

CHAPTER 9

Denouement

September 18, 2018. The law office of Huie, Fernambucq, and Stewart, LLP. Birmingham, AL. The law firm is that of my attorney, Daniel Fortune. It is the day of my deposition. The next few hours will satisfy the conditions of the agreement, and I will leave here a free man. Will Hill Tankersley, Jr. and Michael Taunton from Balch & Bingham are present. Mark DeMichele from Pearson is here along with Pearson Senior Counsel John Garry. So is court reporter Michelle Parvin who will take down the deposition in stenotype. Mr. Fortune, Hunter Carmichael—an attorney from Huie—and my uncle accompanied me into the conference room. Tankersley is all smiles. He approaches me and offers to shake my hand. "I've been waiting to meet you," he says jovially. Taunton is next to him. He is a younger man, thin, dark brown wavy hair, glasses. He's smiling, too. What are they smiling at? They want to pretend that this is a friendly meeting. It is not. We all settle into our seats, and there's some miscellaneous chatter, feigned pleasantries. I notice that Fortune looks even less amused than I am. He knows the firm of Balch & Bingham from the Superfund bribery case. He probably knows more about what happened in that case than he would say. He's seen the tricks these guys have pulled. Ms. Parvin settles in. Tankersley cracks a joke. He thinks he's funny. There are a few polite snickers. My uncle even gives an uneasy laugh. I think I smiled out of awkwardness. I glance at Fortune. He's not laughing. His face is rigid, locked in a scowl sculpted from his years as a government prosecutor.

The Interview

After Fortune stops Tankersley from asking about my job at Synopsys, we transition to my creation of MathLabAnswers. I take them through it, detail by detail, answering honestly, Tankersley's questions and not answering when Fortune interrupts to advise against it. We squabble over a few words that mischaracterize what I did, like, "crack," "invade," and "hack," but for the most part the interview is straightforward. We come to the end of the versions of my program. The questions demonstrate that Pearson's software engineers had no idea what I had done; they reinforced the proof that they never reverse engineered anything. We revisit the time in January 2016 that I emailed Mr. DeMichele asking about Pearson's program. He's nervous.

"You email the people?" Tankersley asks, referring to Pearson's employees. "Right. Yes, sir," I reply.

DeMichele interrupts. "And did I reveal anything?" I can tell he's got a lot riding on my response.

"He didn't," I tell Tankersley.

"He did?" Tankersley asks. DeMichele shifts in his chair. "No, he didn't."

"Did not," DeMichele adds.

"Did not? Did not?" Tankersley repeats, his voice rising. "He did not," I state.

Tankersley cracks a smile. "Now that he's got his heart back and got his—now that his heart is back inside his chest, we'll keep going for a second here." He takes a breath. "Ultimately, nobody at Pearson gave you any information, right?"

I confirmed that no one at Pearson replied to my queries. Tankersley turns to focus on the deletion of data.

"What's your plan about how to get rid of data that's on the Cloud account?" he asks.

"I'll do a complete removal of Google," I tell him.

"That's something I'm not all that familiar with," he admits. "You give a command to remove, and it overwrites it?"

"No. It moves it to a trash section. Once I do the trash deletion it's actually gone."

Tankersley grows concerned. He either doesn't believe me or isn't sure that I'm correct that the data would be permanently gone. The data of my program and what I collected from Pearson is crucial going forward. They don't want any record of what I discovered about Pearson's vulnerabilities and lapses. I had mentioned my internship with Google earlier, so he began to fish for information about how Google goes about deleting information.

"I'm able to do a permanent delete with the Google tool, and there's no way to get it back from my account. But I can't go into the proprietary details of Google," I tell him. "I signed an agreement on that. I can't do that." Frustrated, Tankersley backs off. Later, he decides he wants to confer with DeMichele and tells Ms. Parvin that "this doesn't need to be on the record."

They confer, and the deposition resumes. Fortune has to leave, and Mr. Carmichael takes over as my advising attorney. Tankersley returns and wants to continue discussing the destruction of data.

Mr. Carmichael tries to reassure him. We've been over it already, but Tankersley and Pearson's Senior Counsel John Garry are antsy. Mr. Garry wants a break to go off record. Tankersley, though, wants to hear what Carmichael has to say.

"Working with Mr. Fortune and Mr. Jackson, we'll be sure to have some sort of certification to guarantee that the data is deleted and no longer publicly accessible or available to outside parties," Carmichael advises.

Tankersley jumps in. "The Agreement for Dismissal requires that there be destruction or return of data and software. We would like to have a copy of the software, so we can examine it. But on top of that, sooner or later, you all are going to have to provide some kind of certification to the district attorney that this data has been destroyed. We would want a copy of that certification."

Carmichael, though, objects to Tankersley's demands. "My understanding of the plea agreement is that there's an election whether to return copies or destroy."

"We're making a formal request," Tankersley adds, "that we would like to have a copy of the software just because we want to, you know, maybe look at it."

While I listened to the lawyers' banter, I was thinking about Tankersley's demands for my software. I had created my program, and he and Pearson wanted to take it so they could learn how it worked. And they wanted to learn from it for free. By being allowed to examine my software, they would be getting a free indirect lesson from me on cybersecurity.

I decided to burst his bubble. "I do not have a copy of the first and second versions of my program."

Tankersley looked shocked, scared. I thought he was going to shit a brick. "I thought it was on that CD there," he said, pointing to the table.

"No, no. This is forensic stuff that the UAPD took from my computer. This has the obfuscated versions."

"But what about...didn't they," his voice shook, "give you the laptop back?"

"They did," I replied. But it doesn't have the source-code on it. I removed everything when the third version came out."

"Why did you wipe them?"

"I removed everything before the police came to my room."

He looked exasperated. "How did you know the police were coming?" "I didn't."

"Then why did you take it off?"

Pearson's ploy was to get the police to raid my room and confiscate my devices before I knew what was happening. That way, it could catch everything it wanted from my computers. My deletion beforehand, though, is what had

Tankersley tied in knots. The raid hadn't worked as well as Pearson thought.

"I have a habit of doing that. Once a code doesn't work, I just delete it."

Tankersley is floored. It's clear he's under orders from Pearson to get all the data and versions that I created. I can see him thinking, the wheels turning. I know he thinks that the deleted data is somewhere. So, he returns to the day of my arrest.

"You're working on version three of your program, and then, you get a knock on the door, and it's the police? Where were you when they came into your room?"

Carmichael interjects. "I think this is beyond the scope of my understanding of the terms of what we had outlined..."

"Sure," Tankersley says.

"...as to the interaction with the police," Carmichael concludes.

Then Tankersley unnecessarily feigns innocent curiosity. "Yeah, I'm really just curious about what happened with the laptop, and, you know, whether Mr. Jackson had the bizarre experience where he's watching all these police..."

He keeps going, getting excited, as if he's all caught up in the story when what he wants to know is what the police did, what and when with my laptop. He's got to be able to explain that to Pearson.

"...I wonder what's going on there, him watching..."

Carmichael has enough. This guy, Tankersley, I can tell we're both thinking, is a slime bucket. "I'm sorry," Carmichael interjects again. "I think that's outside of the scope. You can get a police officer..."

"Sure, but he's already testified about what happened," Tankersley says defensively.

"The road we're going down, this is versus what we had outlined."

Now Tankersley interrupts. "I'm just curious about the circumstances under which the laptop was taken. I'm entitled to know that," he declares petulantly.

The laptop and its handling is haunting him. He wants to know what the police did with it, and to it, when they had it, but I've already told him I didn't know. Tankersley gives up and decides he has to be content with my attorney, Mr. Fortune, taking possession of it.

Tankersley is talking in circles now. He comes back to Google.

"Your plan is to direct Google to destroy or make inaccessible the data on their end, right?"

"Yes, sir," I assure him.

"And you can't tell me how they did it because I'm not entitled to know and you're under an agreement not to tell me, right?"

"Right." This pisses him off. He thinks, of course, he's entitled to it.

Mr. Garry says he wants to tell Mr. Carmichael something off the record. This Garry guy—Pearson's in-house attorney—always wants to run his mouth off the record. Before talking to Carmichael, he kept interrupting the deposition and demanding to go off the record, so he and DeMichele could accuse me of stealing from Pearson. He didn't want that accusation on the record. Coward.

"You guys made the site publicly accessible," I shot back. Plus, I couldn't understand why they wanted to continue to accuse me of crimes when they had agreed to this deposition. This wasn't the forum for this. These guys couldn't help but be unprofessional. It's clear they were still scared of what I knew. And angry. They were losing their composure.

Then, Tankersley wants off the record, too. He wants to know what I did with all that student data that Pearson had left exposed.

"I gave it to my attorneys. Daniel Fortune and David Schoen," I told him. "What was that? You gave it to whom? Mr. Fortune and...?"

"David Schoen." I think he shit another brick. These guys at Balch know Schoen's reputation as a bulldog civil attorney.

Tankersley takes a deep, measured breath. "You gave David Schoen," he proceeds slowly, "you gave David Schoen all that data?"

"Yes," I answered.

His neck seemed to tighten right in front of me. He sighed. "Ok, ok. Let's move on," he said, gathering himself. I could see him pondering what Schoen might do knowing that Schoen now knew and had evidence that Pearson Education may have violated major FERPA laws. All it would take is for Schoen to file a complaint with the Department of Education on behalf of perhaps all the students whose information Pearson left exposed on publicly accessible sites and programs. We're talking about millions of students.

It's time to wrap it up. Mr. Taunton has not said a word. Tankersley's aggression reinforces my suspicion that he's behind the fake EULA they tried to pull. Shakedown.

"You've got the Agreement for Dismissal in front of you," Tankersley says. "I take it that you don't have any plan to access any Pearson data now or in the future is that correct?"

"Yes, sir," I tell him.

"You don't have any plans to create any new players that are directed at Pearson software?"

"No, sir." Just sign the agreement. This guy.

"You don't have any plans to access anymore Pearson user data?" His condescending tone was eking through my bones. He was talking to me like I was seven.

"No, sir."

"And you don't have any plans to use any of the Pearson user data that you have at this time?" His face, the smugness.

"No, sir, I'm not going to use it." I was being polite because I just wanted to get out of there.

"And you don't have any plans to use any of the software that you've created relative to your MathLab products that you have at this time?"

"No, sir."

"Is there anything else about the Pearson data that you've not already testified about?"

"No, sir."

"I don't have any further questions."

Afterward, these Pearson and Balch guys—Garry, Tankersley, Taunton—had the gall to come up to me and shake my hand, telling me, "Use your powers for good." They have no shame.

"Thank you, Mr. Jackson," Tankersley adds. "I hope you have a bright future ahead of you." At least, this time, he got my name right.

CHAPTER 10

Endgame

June 2019: David Schoen sends an email to Daniel Fortune and Joel Sogol. My case was finalized the previous October. Pearson agreed that I had met the terms of the agreement during the deposition. All criminal charges were dismissed. Schoen and I, however, still had some concerns. After all that had transpired, we weren't sure that we'd heard the last of Pearson Education. I knew that there might have been some fallout from my case. At some point, UA's math instructors stopped using Pearson's products for their math courses. I heard that other UA instructors stopped using Pearson's MyLab software as well. Although my MathLab-Answers program mainly targeted math students, it could be used for any of Pearson's MyLab courses. I couldn't determine if my case had anything to do with the University no longer using Pearson MyLab, but something tells me that it may have played a part. Schoen and I were concerned that Pearson might be watching me to see if it could find anything to accuse me of violating the agreement. Pearson had monitored people before, so what was to stop them from doing surveillance? The corporation had done extensive research on me. It had my name and address, of course, and my emails, usernames, a list of my current and past employers, and the names of my relatives. There was also no telling what its lawyers from Balch might try to pull.

Returning Fire?

Schoen and I considered bringing a lawsuit of our own against Pearson, and we wanted to gauge Fortune's and Sogol's thoughts on the matter. Schoen and I were indignant over what had transpired over the last three years, how Pearson had upended my life, affected my schoolwork, and, yes, that little matter of almost railroading me into prison. Schoen's experience in fighting for people's civil rights had led him to see much injustice and corruption over the years, so he had a pugnacious attitude in him to make him want to fight for me and hold Pearson Education accountable. In addition to contemplating suing Pearson, Schoen wanted Fortune's and Sogol's advice about reporting Pearson to the Department of Education for violation of FERPA laws. For Schoen, the latter would hold them accountable to the public for leaving millions of students' information out in the public domain. But Fortune and Sogol believed I just needed to sit tight for a while. I needed to get through school. Remember, I'd been a college student this whole time trying to get through my coursework while fighting this legal battle. I was set to graduate in December 2019.

Since signing the agreement, I've been in good spirits. I no longer had to worry about prison or having a felony conviction on my record. A civil suit from Pearson didn't scare me. My grandmother's cancer, though, had worsened and my grandfather had some health problems as well. Plus, I was still haunted by the experience and felt that I hadn't gotten justice. These guys—Pearson and Balch & Bingham—had tried to destroy my career and my life. My girlfriend and my mom thought I might want to see a counselor to help me through some of my anger. But, to me, the best therapy is justice.

Justice can come in different forms, and over the years, I'd been meeting with my former English teacher Dr. Love to discuss the case, and we began to discuss the possibility of a book. I liked the idea because it would be a way for me to get my story out; I wanted the public to know what this "education" company had done to me. But he had warned me that he couldn't advise writing a book until the criminal case had come to an end and I received approval from my lawyers. Now that the case was closed, and with the greenlight from my attorneys, we began to seriously discuss the writing of the book. So, we met periodically for lunch, and I was able to talk about all that I had been through with Dr. Love. In a way, these meetings were like therapy and knowing that I could write my story improved my outlook.

Back To Normal

By the summer of 2019, I started to return to my normal life, having fun, working, and doing stuff I enjoyed. College is supposed to be one of the most exciting and fun times in a young person's life, but because of the Pearson case, I hadn't been able to enjoy much of my time at Alabama. In June, though, I won a cruise to Key West competing in a Rock, Paper, Scissors tournament in Gulf Shores. In the tournament, I never lost a game. I was also competing in video game tournaments in Daytona, Dallas, Canada, and Europe. I was a semi professional in a 2D fighter game. Not only did I play the game, I conducted vulnerability assessments for the game's tournament partner StartGG. It turns out, StartGG had some vulnerabilities in its system, so I'm working to make sure that the company's tournament platforms are safe. In exchange, I get to play in the tournaments and travel the world.

And this—me working with companies—is what irks me about Pearson. Instead of having me arrested, all Pearson had to do was send me an email or correspond with me in a professional forum, and I could have helped the company with its vulnerabilities. If its executives didn't like my website, its representatives and lawyers could have discussed with me what they thought was either illegal or problematic, and we could have worked out a resolution. Instead, they sicked the police on me before I even knew what was happening. One of the projects I'm developing is building a platform for cybersecurity experts and legal hackers and companies to communicate and work together to address vulnerability issues with companies' online presence and programs. But I'm sure I know why Pearson took the strategy it did. Pearson and Balch & Bingham saw I was just some kid from Alabama, a black kid to boot, and they thought they'd scare the shit out of me, and I'd roll over and hand over, tell them, and do everything they wanted.

I also started to work on independent and innovative cybersecurity programs that I planned to market to different companies and to secure a job after graduation. I created a program called SADD—Scalable Anonymous Disposable

Desktop—which has a patent pending. The details of the program are posted at https://sadd.io. It's a program that allows you to operate a virtual computer on your desktop, laptop, and phone that's anonymous and can permanently delete all data—the ultimate in online cybersecurity. In June 2019, I traveled to Huntsville, AL to present SADD where I met with Department of Defense contractors about one of my new programs. They were so impressed with the program that they made me a job offer. I had, of course, to go through a thorough background check and tell them all about my ordeal with Pearson and my hacker past when I was a kid. Rather than see these as negatives, the DoD went ahead with its plans to hire me as a reverse engineer. In the fall of 2020, I earned my Top Secret security clearance.

Collectively, we don't realize how vulnerable our information is. Even large corporations have trouble seeing their vulnerabilities despite hiring people who are loaded up with degrees and experience. They pay these experts a ton of money to safeguard the company's information and the information that companies get from the public. Some companies think that because they've invested millions into cybersecurity, hired whom they think are the best engineers they can find, and have yet to experience a breach that they're safe. But they're not.

To protect their identities, I won't give names, but in one of my business ventures, I got in contact with an official from a nationwide chain of casinos. I advised him that the casinos' systems were vulnerable to infiltration and attack. A smart criminal hacker could steal millions, and they would never even know what happened. Arrogant, he didn't believe me.

He said, "Not us. We're impenetrable." I told him I could prove it and asked him if he wanted me to test them. "Sure, go ahead," he replied.

Not long afterward, I called him up. I can't reveal what I saw, but I named off item by item what I was seeing.

He paused. Then, he said, "Oh shit."

If casinos aren't secure protecting their own money, imagine how unsecure companies' systems are that handle the public's money and private information. While my case was unfolding, I also gave a demonstration to the Hayneville Police Department showing the vulnerabilities in the state's law enforcement online programs. My hope is to help the state find ways to suture these holes in its software.

Next Steps

Contrary to how Pearson Education and Balch & Bingham tried to make me look, ever since I started to become an adult, I knew that I would create legitimate programs and platforms that would help individuals and businesses develop more secure networks, software, and online presence. Just as importantly, as a kid who came from Lowndes County and the descendant of civil rights leaders who risked their lives and livelihood to make people's lives and their communities better, I want to use my knowledge and skills to help kids like me or who have it worse than I ever did. In December 2020, my grandmother passed away from cancer. Remembering her life and all she stood for, I realized how fortunate I was to have a strong family and supportive ex-girlfriend, although our relationship ended. I was fortunate to have the attorneys I had. Many of the kids and young people I know and grew up around don't have the resources

I had, and if they ever got targeted by a company like Pearson or by a relentless district attorney, they'd have no chance. I visit and talk to kids in afterschool programs who are stuck in poverty, struggling schools, and broken families, and I talk to them about how important it is to find something they like to do and that they're good at. When they hear it from someone who looks like them, can talk like them, and can relate to them, they start to believe in themselves. I can't promise them that nothing bad will ever happen or that they won't be accused of something they didn't do. But I can at least give them the right skills to fight back against whatever it is that they face.

I think about all that my family has been through and what they have accomplished, and I know this: my story isn't just my story. Any story worth telling is a story about more than just one person. Just like the story of my grandfather fighting for civil rights isn't just about him but what he did for others and what everyone like him did for each other. I'd like my story to be like my grandparents' story: a story that shows that whether it's against the Klan, a shady law firm, or a billion-dollar corporation, you never have to stop fighting for what's right. Like my grandfather once said, education and enlightenment are the best tools to fight injustice. And part of that enlightenment and education comes from the stories we tell to and share with each other—whether they're stories from 1960s Lowndes County, Alabama or from the campus of a large university in the 21st century. By sharing them, we can be proud of what's changed but still on guard for what's to come. And I'll be near my computer, not waiting, but preparing, learning, and teaching, because that's what my grandparents would want me to do.

The End

Printed in the USA
CPSIA information can be obtained
at www.ICGtesting.com
LVHW031501270824
789423LV00016B/155/J

9 780979 543456